# PRINCIPLES OF ANIMAL DESIGN

C. Richard Taylor on July 5, 1995, with friends in the hills near Ascona.

# PRINCIPLES OF ANIMAL DESIGN

## The Optimization and Symmorphosis Debate

*Edited by*

EWALD R. WEIBEL
*University of Berne*

C. RICHARD TAYLOR
*formerly of Harvard University*

*and*

LIANA BOLIS
*University of Milan*

CAMBRIDGE
UNIVERSITY PRESS

Published by the Press Syndicate of the University of Cambridge
The Pitt Building, Trumpington Street, Cambridge CB2 1RP, United Kingdom

Cambridge University Press
The Edinburgh Building, Cambridge CB2 2RU, United Kingdom
40 West 20th Street, New York, NY 10011-4211, USA
10 Stamford Road, Oakleigh, Melbourne 3166, Australia

First published 1998

Printed in the United States of America

Typeset in Times Roman by KW

*Library of Congress Cataloging-in-Publication Data*
Principles of animal design: the optimization and symmorphosis debate
/edited by Ewald R. Weibel, C. Richard Taylor, Liana Bolis.
p.  cm.
Papers from the Twelfth International Conference on Comparative
Physiology, held July 1–5, 1995, on Monte Verità above Ascona,
Switzerland.
ISBN 0-521-58370-5 (hardcover). – ISBN 0-521-58667-4 (pbk.)
1. Morphology (Animals) – Philosophy – Congresses.  2. Adaptation
(Biology) – Congresses.  3. Evolution (Biology) – Congresses.
4. Physiology – Congresses.  5. Symmorphosis – Congresses.  I. Weibel,
Ewald R.  II. Taylor, C. Richard (Charles Richard).
III. Bolis, Liana.  IV. International Conference on Comparative
Physiology (12th: 1995: Ascona, Switzerland)
QL799.P6635  1998
571.3'1 – dc21    97-13547   CIP

*A catalog record for this book is available from the British Library*

ISBN 0 521 58370 5 hardback
ISBN 0 521 58667 4 paperback

# Contents

# Contributors

R. McNeill Alexander
*Department of Pure and Applied Biology, University of Leeds, Leeds LS2 9JT, UK*

Andrew A. Biewener
*Department of Organismal Biology and Anatomy, University of Chicago, 1027 East 57th Street, Chicago, IL 60637, USA*

Mohamed Bnouham
*Départment de Biologie, Faculté des Sciences, Université Mohamed Ier, Oujda, Maroc*

Liana Bolis
*Laboratory of General and Comparative Biology, University of Milan, Via Balzaretti 9, I-20133 Milan, Italy*

Kevin E. Conley
*Magnetic Resonance Laboratory, SB-05, Department of Radiology, Box 357115, University of Washington Medical Center, Seattle, WA 98195-7115, USA*

Alfred W. Crompton
*Museum of Comparative Zoology, Harvard University, 26 Oxford Street, Cambridge, MA 02138, USA*

Pierre Dejours
*Centre National de la Recherche Scientifique, Centre d'Ecologie et Physiologie Energétiques, 23 rue Becquerel, F-67087 Strasbourg, France*

Jared M. Diamond
*Department of Physiology, UCLA School of Medicine, 10833 Le Conte Avenue, Rm. 53-246 CHS, Los Angeles, CA 90024-1751, USA*

Pieter Dullemeijer
*Institute of Evolutionary and Ecological Sciences, Theoretical Biology Division, Kaiserstraat 63, NL-2300 RA Leiden, The Netherlands*

Martin E. Feder
*Department of Organismal Biology and Anatomy, University of Chicago, 1027 East 57th Street, Chicago, IL 60637, USA*

Theodore Garland, Jr.
*Department of Zoology, University of Wisconsin, 430 Lincoln Drive, Madison, WI 53706, USA*

James W. Glasheen
*5130 Valley Life Sciences Building, University of California Berkeley, Berkeley, CA 94720, USA*

Malcolm S. Gordon
*Department of Biology, University of California, Los Angeles, CA 90095-1606, USA*

Kimberly A. Hammond
*Department of Biology, University of California, Riverside, CA 92521, USA*

Peter W. Hochachka
*Department of Zoology, University of British Columbia, 6270 University Boulevard, Vancouver, BC, Canada V6T 1Z4*

Reinhold R. Hofmann
*Institut für Zoo- und Wildtierforschung (IZW), Alfred-Kowalke-Strasse 17, D-10315 Berlin, Germany*

Hans Hoppeler
*Department of Anatomy, University of Berne, Bühlstrasse 26, CH-3000 Berne 9, Switzerland*

Connie C. W. Hsia
*Department of Internal Medicine, University of Texas Southwestern Medical Center, 5323 Harry Hines Boulevard, Dallas, TX 75235-9034, USA*

Ian D. Hume
*Biological Sciences A08, University of Sydney, Zoology Building, Sydney, NSW 2006, Australia*

James H. Jones
*Department of Surgical and Radiological Sciences, School of Veterinary Medicine, University of California, Davis, CA 95616-8732, USA*

Richard D. Keynes
*Physiological Laboratory, University of Cambridge, Downing Street, Cambridge CB2 3EG, UK*

Simon B. Laughlin
*Department of Zoology, University of Cambridge, Downing Street, Cambridge CB2 3EJ, UK*

Yvon Le Maho
*Centre National de la Recherche Scientifique, Centre d'Ecologie et Physiologie Energétiques, 23 rue Becquerel, F-67087 Strasbourg, France*

Daniel E. Lieberman
*Department of Anthropology, Rutgers University, PO Box 270, New Brunswick, NJ 08903-0270, USA*

Stan L. Lindstedt
*Department of Biological Sciences, Northern Arizona University, Flagstaff, AZ 86011-5640, USA*

Simon Maddrell
*Department of Zoology, University of Cambridge, Downing Street, Cambridge CB2 3EJ, UK*

John N. Maina
*Department of Anatomical Sciences, University of the Witwatersrand, Medical School, Parktown 2193, Johannesburg, South Africa*

Ricardo Martinez-Ruiz
*Departamento de Fisiologia, Instituto de Medicina Experimental, Universidad Central de Venezuela, Caracas, Venezuela 1050*

Eviatar Nevo
*Institute of Evolution, Haifa University, Mount Carmel, Haifa 31905, Israel*

Dirk Pette
*Fakultät für Biologie, Universität Konstanz, PO Box 5560 M641, D-78434 Konstanz, Germany*

David Randall
*Department of Zoology, University of British Columbia, 6270 University Boulevard, Vancouver, BC, Canada V6T 1Z4*

Robert E. Ricklefs
*College of Arts and Sciences, Department of Biology, University of Missouri, 8001 Natural Bridge Road, St Louis, MO 63121-4499, USA*

Thomas J. Roberts
*Department of Biology, Northeastern University, 414 Mugar Building, Boston, MA 02115, USA*

Lawrence C. Rome
*Department of Biology, University of Pennsylvania School of Arts and Sciences, Leidy Laboratory of Biology, Philadelphia, PA 19104-6018, USA*

Knut Schmidt-Nielsen
*Department of Zoology, Duke University, Box 90325, Durham, NC 27708-0325, USA*

Amiram Shkolnik
*Department of Zoology, Tel Aviv University, Ramat-Aviv, 69978 Tel-Aviv, Israel*

Paul A. Srere
*Department of Veterans Affairs Medical Center, 4500 S. Lancaster Road, Dallas, TX 75216, USA*

Robert S. Staron
*College of Osteopathic Medicine, Ohio University, Irvine Hall, Athens, Ohio 45701-2979, USA*

Raul K. Suarez
*Department of Ecology, Evolution, and Marine Biology, University of California at Santa Barbara, Santa Barbara, CA 93106-9610, USA*

H. Lee Sweeney
*Department of Physiology, University of Pennsylvania School of Medicine, 3700 Hamilton Walk, Philadelphia, PA 19104-6085, USA*

C. Richard Taylor
*Concord Field Station, Museum of Comparative Zoology, Harvard University, Old Causeway Road, Bedford, MA 01730, USA (deceased September 10, 1995)*

Edwin W. Taylor
*School of Biological Sciences, University of Birmingham, Edgbaston, Birmingham B15 2TT, UK*

Steven Vogel
*Department of Zoology, Duke University, Box 90325, Durham, NC 27708-0325, USA*

Jean-Michel Weber
*Department of Biology, University of Ottawa, 30 Marie Curie Street, PO Box 450, Station A, Ottawa, Ontario, Canada K1N 6N5*

Ewald R. Weibel
*Institute of Anatomy, University of Berne, M. E. Müller Foundation, PO Box 620, CH-3000 Berne 9, Switzerland*

Daniel Weihs
*TECHNION – Israel Institute of Technology, The Graduate School, Technion City, Haifa 32000, Israel*

Wolfgang Wieser
*Institut für Zoologie und Limnologie der Leopold Franzens Universität Innsbruck, Technikerstrasse 25, A-6020 Innsbruck, Austria*

# Preface

The Twelfth International Conference on Comparative Physiology was held on July 1–5, 1995, at the historical conference site on Monte Verità above Ascona in the southern part of Switzerland. It was conceived and organized in collaboration with the Interunion Commission for Comparative Physiology, set up by the International Unions of Physiological Sciences (IUPS) and of Pure and Applied Biophysics (IUPAB), led by Professor Malcolm S. Gordon who had succeeded C. Richard Taylor as its chairman in 1993.

The topic of optimization in biological design and of symmorphosis is highly controversial and so it seemed fitting to attempt a "search for the truth" by assembling on Monte Verità – the Mountain of Truth – a group of biologists from many different disciplines, true to the tradition of these conferences which, on purpose, have always stressed the interdisciplinary intercourse over too specialized reductionistic exercises. The discussions were fascinating as they extended from molecular biology and biochemistry of enzyme systems through the study of bone and muscle to integrative systems of energy supply and, to a limited extent, to the nervous system. The points of view were those of physiology and morphology on the one hand, and those of evolutionary biology on the other hand. The authors have made a special effort to present the essays based on the presentations in a form that should be accessible by a broad readership of biologists interested in basic principles. We hope that they will challenge thoughts and perhaps even new research on the principles by which organisms are made.

The conference was made possible by generous financial contributions from Ciba AG, F. Hoffmann-LaRoche AG, and Sandoz Pharma AG, all in Basel, the Maurice E. Müller Foundation for Research and Continuing Education in Orthopedic Surgery in Berne, the Swiss

Academy of Medical Sciences, the Swiss Academy of Sciences, the Swiss National Science Foundation, and the Union of Swiss Societies for Experimental Biology. We are most grateful for this support.

We would also like to thank the staff of Monte Verità for their great efforts on behalf of the conference and the participants, Drs Ruth Vock and Hans-Rudolf Widmer for their collaboration in the organization, Ms Elsbeth Hanger for her skillful assistance in the editorial work, Dr Hans Howald for the preparation of the index, and Dr Robin Smith from Cambridge University Press in New York for an excellent and helpful collaboration in producing a handsome volume.

This preface has two additional sections, for very special reasons. The first is a memorial tribute to the long-term Chairman of the Interunion Commission for Comparative Physiology, and by that the leader of many of the Conferences on Comparative Physiology, C. Richard Taylor, who sadly died two months after this conference. With this tribute we express our deep gratitude for all he has done to foster comparative physiology.

The second is a special article related to the origins of the Conferences on Comparative Physiology that have been run, over the past quarter of a century, by a few people. The foremost and most persistent driving force was Liana Bolis, Professor of General and Comparative Physiology at the University of Milan in Italy, who started these conferences in 1970 and assumed the main responsibility for their organization ever since. For this reason C. Richard Taylor invited Professor Bolis to speak and write about "her life with animals" that had prompted her to learn about biology. With its compassion and emotional touch this little narrative complements beautifully the (perhaps overly) mechanistic essays on animal design that follow.

*E. R. W.*

# In memory of
# Charles Richard Taylor (1939–1995)

*Ewald R. Weibel and Liana Bolis*

The Symposium on "Optimization in Biological Design: Controversies about Symmorphosis" took place early in July 1995 in Ascona in Switzerland as the Twelfth International Conference on Comparative Physiology. Dick Taylor had lent all his force to the conception and organization of this symposium, as he had done for the preceding conferences over nearly two decades. He participated in this conference in full vigour, with challenging remarks in the discussions, and, at the end, with a brilliant extemporaneous summary of the proceedings that was as charming as it was thoughtful.

While we were preparing for editing the papers in this volume he succumbed, on September 10, 1995, at the age of 56, to his long-standing illness which he had never allowed to slow him down.

Charles Richard Taylor, born on September 8, 1939, was the Charles P. Lyman Professor of Biology at Harvard University. One of the leading integrative physiologists, he strived to unravel the intricacies of how animals work by looking at them as a whole. In 25 years as Director of the Concord Field Station of Harvard's Museum of Comparative Zoology he developed a unique research strategy oriented on systems physiology, and he tackled it mainly with the tools of comparative physiology, exploiting the diversity in nature to discover unifying principles of biological design.

Together with his exceptional drive and charisma this made him the natural leader in the International Conferences on Comparative Physiology, taking over from the early leadership of his teacher Knut Schmidt-Nielsen. He became the chairman of the Interunion Commission for Comparative Physiology that was instituted jointly by the International Union of Physiological Sciences (IUPS) and the International Union of Pure and Applied Biophysics (IUPAB); a function he retained until 1993 when his mandate expired. Under his leadership the activities of this Commission, and mainly these International Conferences, have been exceptionally successful and interesting, largely because of the extraordinary scope they covered.

It was sad, but in a way rewarding, that his last conference should be on a topic that was particularly close to his heart and on which he has spent so much of his efforts and thoughts over the past two decades: the question of whether animals are well designed to perform their function. This book is something like a legacy of the scientist Dick Taylor.

The origin of the concept of symmorphosis can be traced back to the first paper on the design of the respiratory system that resulted from the joint work between his laboratory at Harvard University and the Department of Anatomy at the University of Berne in Switzerland. In this paper Dick had entered the sentence: "In undertaking this task we were motivated by the firm belief that animals are built reasonably" and from this then derived the principle of symmorphosis which postulates that the quantity of structure an animal builds into a functional system is matched to what is needed. This notion was exciting, first because it was a great challenge to see whether this is really the way nature works and where the principle breaks down, and second because this bold concept provoked a controversy that culminated in this symposium.

With this background it was only logical that Dick Taylor also believed that animals make the best possible use of the equipment they have. For example, horses use their legs differently when they run at low speed than when they gallop, and Taylor and his collaborators showed that the transition between gaits occurs at the speed where it becomes more economical to either gallop or to run. Thus he always asked the question of functional economy, including when he observed how African women carry their loads on their head with great ease and at almost no energetic cost.

The study of complex functional systems and their design is not an easy matter. Taylor's strategy was to first ask what could be learned from looking simply at how nature has solved its problems when designing animals of different capacities for a variety of reasons, be it the variation in size, from the mouse to the elephant, or because some animals are more athletic than others. Dick has led his group to make the best use of the vast variation observed in nature to unravel the secrets of animal design. And he could build not only on his broad knowledge and experience as a naturalist, but just as much on his unusual skills in working humanely with animals in testing their performance; horses, lions, or elephants worked with him just as willingly as his collaborators.

But no one man, not even one of the stature of a Richard Taylor, can carry out the study of complex systems all by himself. It is probably Dick Taylor's greatest achievement that he succeeded in bringing together into joint interdisciplinary projects a whole army of the best scientists from America to Europe, Israel, Africa, and Australia, to tackle the problems from very different perspectives, and then to synthesize the results into a coherent message.

We deplore the loss of Dick Taylor not only as an exceptionally creative and outstanding scientist, but just as much as a dear friend. We feel honored for having had the privilege to work with him almost since the foundation of the International Conferences on Comparative Physiology in the early 1970s and in the Interunion Commission. Dick Taylor perfectly combined zoology and experimental physiology in the spirit of the Aristotelian concept, and through the International Conferences he opened new ways and new views to the scientific community, fostered by his rigorous work and broad thinking. In the early 1970s, comparative physiology was deeply rooted in comparative anatomy. Following the path laid by Knut Schmidt-Nielsen, Dick Taylor, his brilliant student, gave this field the quality of an independent discipline that can form the junction between zoology, physiology, and anatomy. Through these conferences he also succeeded in spreading to the broader scientific community what he gave to his students, collaborators, and friends: the compassion for precise scientific work in a broad perspective, a facility of becoming excited about an interesting observation, and a great generosity and humane kindness.

# A life with animals: from cat to fish

*Liana Bolis*

Animals are fantastic creatures, very much linked to human beings. Apart from some contrasts related to the transmission of diseases or aggressive behavior, animals are friends with man. Like man, animals, together with plants, are part of the universe. But what is particularly interesting is that man needs animals, whereas animals do not need man.

If we read about the origin of the universe in the book of Genesis, we see that during the Creation the heavens and the earth were formed. The earth was without any living form and was a desert, but later water and land were divided, night and day were formed with the different lights of the moon and the sun, and from the water came Life. Later came birds and reptiles and other animals; and finally, man and woman. According to the Bible, the Creation lasted seven days. These seven days may represent billions of years for us, as explained by the theory of evolution. Evolution and adaptation interacted to finalize the world as it is.

Through domestication, animals became increasingly submitted to man and a different pattern of behavior between man and animals developed. It is well known that not only children love animals: the mystery of communication linked only by subtle perception is sometimes intriguing. With cats and dogs a nonverbal communication has developed, which brings man closer to animals.

As a child I loved animals and particularly cats, which I could keep in my house. Dolls were too obedient for me and I preferred to interact in my play with a living creature. The cat was obedient until a certain time and after my lovely pussy-cat disappeared my heart was full of sorrow.

At that time I had a very interesting book at home: *The Private and Public Life of Animals* (Grandjean), edited in 1824. That book was particularly fascinating since the author reversed the relationship between man and animals showing how they could adopt some human policies. In

xviii

the book, different kinds of animals developed meetings to discuss problems of common interest, where the cat was always present as secretary. Marriage between a butterfly and a dragonfly was very well discussed, with the main aspect of the marriage being the protection of the wife by the husband! A nice story of a hare related the problems of hunting combined with protection of the babies while the female rested. Another tale told of the bear dreaming of summer life during winter. In reality we try to understand how animals live, but we should also think that animals might be interested in mimicking our lives.

Later, when I had obtained my medical degree, I was still very interested in animals. I did not like performing experiments with rats and mice but thought to turn to aquatic life; so fish were an important alternative in my scientific life. I went backwards in evolution: from cat to fish.

My ultimate happiness came from an invitation to participate in a marine biological expedition: From Tromso to the Coral Sea, from the Gulf of Mexico to Alaska and to the Amazon. Observing the beautiful water life was truly fascinating; there could not have been a better chance to study, observe, and carry out experiments in such a fascinating setting of the real sea life.

But the most interesting marine expeditions were always in the waters around Cormorant Island, with the lovely village of Alert Bay, in the North Pacific of British Columbia. It is a small island situated at the end of the channel between the mainland and Vancouver Island. We had several expeditions there with the prestigious boat, the *Alfa Helix* (a marvellous boat with all facilities for carrying out experiments offshore), or other boats. I went there every August and September from 1970 to 1989 when we no longer had a boat – but the inhabitants of the small village of Alert Bay prepared a little laboratory for our group to carry out research on a collection of samples, to be examined in different universities.

The location of Cormorant Island is important since the salmon pass through this area on their journey from the open ocean to the river where they were born. It is a fascinating period: thousands of salmon following the "genetic clock" are heading for the river, almost religiously with a mystic and determined attitude. They do not eat, and they change the color of their skin. They do not miss "their river" and if there is not enough water in the river they begin a ritual "dance" to call for water. They will return to the river to give birth and to die. Death is part of life in their species.

This life cycle of the salmon has dramatically influenced the culture and artistic expression of the Indians, who principally inhabit that area. The Indian fishermen have a very interesting interaction with the salmon. It seems that they thank the fish for giving them enough food and money, and are extremely respectful of the salmon's power over their lives.

Our experiments were exciting: from brain research to calcium metabolism and osmoregulation. The memories of life with the salmon are very rewarding. We are always longing for our short time in Alert Bay.

So the cat went down to the fish evolution, but the importance for our origins remains from philosopher to fish.

# 1

# Symmorphosis and optimization of biological design: introduction and questions

## EWALD R. WEIBEL

Are animals built economically? This seemingly simple question led to the discussions summarized in this volume. Phrased differently it asks whether structures are used sparingly to match functional demand with the result that the *design* of biological organisms would tend to be *optimized* for the functions performed. As reasonable as it seems, the contention that animals should be designed according to strict bioengineering principles raises considerable criticism.

However, there is much evidence that the diversity of species, as it has resulted from evolution by natural selection, represents variations in design suggestive of adaptation to specific tasks, and also of an economic use of resources. Even at the basic level of biological organization, in the cell, economy appears to prevail: although all cells contain the entire genome in their nucleus, only the very small part relevant for the cell's specific functions is expressed. It appears to be a fundamental principle not to load cells with baggage they do not need. But we will have to ask whether the enzyme systems that a particular cell expresses are also *quantitatively* adjusted to the cell's functional needs or whether many of them occur in vast excess.

At the higher levels of biological organization a well-known example of economic design is bone tissue whose fine structure is quantitatively adjusted to the mechanical stresses imposed. This translates into the arrangement of bone trabeculae according to stress lines and into quantitative aspects of the macroscopic design of whole bones such that, for example, the bones of the serving arm of a tennis player are thicker than those of the other arm. Similar effects are known for the quantitative design of musculature where exercise training causes the active muscles to become larger and stronger. Similar differences are found in nature when comparing species that specialize for certain locomotor tasks. Although

1

the muscle mass of mammals is designed on a basic master plan it shows quantitative differences according to need, as shown by the two cases illustrated in Figure 1.1: the Pronghorn antelope of the Rocky Mountains, a champion high-speed endurance runner, increases the muscle mass at the proximal hind limb with which it achieves the main forward thrust; in contrast, the fossorial mole rat reinforces the muscles of the front limb and neck with which it digs its burrows, and it reduces hind limb musculature which it uses only for balance.

A further well worked-out case is the design of blood vessels where the progression of branch diameters along the vascular trees is such as to minimize both the mass of blood needed to fill the vessels and the energy required to transport blood to the targets against flow resistance. The vasculature also has to ensure that all points in the body are evenly served, a problem that becomes particularly evident in the lung where blood flow must be evenly distributed over a gas exchange area of the size of a tennis court. The solution found by nature is to design the vasculature as fractal trees and adjusting the proportionality factor from one branch generation to the next so as to minimize the cost of transport.

Such case studies suggest that the design of cells, of tissues and of organs is adjusted to functional needs, and that the body uses the materials and the space it has available sparingly. Conversely, if the matching of

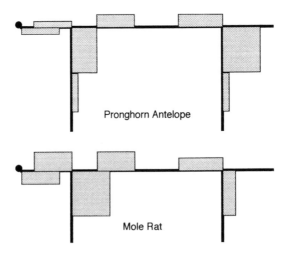

Figure 1.1. Scale models showing the distribution of muscle mass in the Pronghorn antelope of the Rocky Mountains and of the mole rat from Israel (see Chapter 10.4) in accordance with different requirements of locomotion.

structures to functional needs is combined with strict economy then this implies that the design of biological structures must tend to be optimized for the specific function they serve – a strong statement that must trigger fundamental skepticism.

Optimal design implies perfect matching of structure to functional need, both qualitatively and quantitatively. If design is optimal in the sense that there is no more structure than is needed to fulfill the function, then the structural feature becomes the key factor that sets the limit of functional performance, the functional capacity. The degree of optimality in design can then be tested by estimating how the functional capacity is varied with variations of functional demand.

On the other hand, engineers build their systems with some safety factors that make sure that the system does not "break" when the presumed maximal performance is reached. From an engineering point of view this is "good design" as it considers not only the precise function to be performed but also the risks – to the organism – of exploiting the capacity to the limit. This may, however, be a serious impediment to testing whether optimized design prevails because "safety factors" must appear as excess capacity of which it is hard to say whether it is not simply a wasteful use of resources – unless one can define the potential excess load that must be absorbed by the excess capacity.

### Symmorphosis as a hypothesis of optimized integral design

The problems we met in the foregoing discussion resulted largely from focusing on engineering aspects of the design of particular structures. It may, however, be more useful to consider the design of cells and organs in the perspective of the functional systems they serve, rather than as "stand-alone" features. The question then is whether the design of the parts is – quantitatively – adjusted to the overall functional task; in other words, whether the parts are all coadjusted to their common role. It is in view of such questions that we have proposed the hypothesis of *symmorphosis*, which we defined as the "state of structural design commensurate to functional needs resulting from regulated morphogenesis, whereby the formation of structural elements is regulated to satisfy but not exceed the requirements of the functional system" (Taylor and Weibel 1981). This hypothesis has stirred considerable criticism.

Symmorphosis combines three elements: functional performance, functional capacity, and economy of structural design. The hypothesis postulates that capacity is adjusted by design to the (expected) loads on the

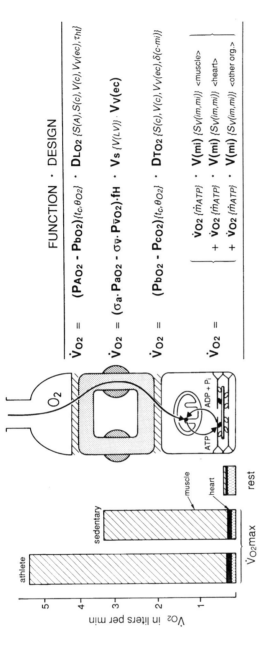

Figure 1.2. Model of the respiratory system. The flow rate $\dot{V}_{O_2}$ is expressed as the product of functional and design parameters shown in bold face; parameters that affect the factors are shown in italics and placed in braces {.}. The functional parameters include $O_2$ partial pressures ($P_{O_2}$), coefficients of "hematocrit-specific" $O_2$ capacitance ($\sigma$) which depend on $O_2$–hemoglobin dissociation, $O_2$ binding rate ($\theta$), heart frequency ($f_H$), capillary transit time ($t_c$), and mitochondrial $O_2$ consumption rate as function of ATP flux $\dot{v}_{O_2}\{\dot{m}_{ATP}\}$. Design parameters include diffusion conductances ($D$) of lung and tissue gas exchangers which depend on alveolar and capillary exchange surface areas ($S(A)$, $S(c)$), capillary volumes ($V(c)$), hematocrit ($V_V(ec)$), harmonic mean barrier thickness ($\tau_{ht}$), capillary-mitochondrial diffusion distance ($\delta(c,m)$), and mitochondrial volume ($V(mt)$)) with inner membrane surface density ($S_V(im,m)$). The histogram to the left shows the distribution of $\dot{V}_{O_2}$ to locomotor muscle, heart and other tissues under resting conditions and at $\dot{V}_{O_2,max}$ in sedentary and athletic humans. (Reprinted from *Respiration Physiology*, **87**, E. R. Weibel, C. R. Taylor and H. Hoppeler. Variations in function and design: testing symmorphosis in the respiratory system, 325–48 (1992) with kind permission of Elsevier Science BV, Amsterdam.)

functional system, and therefore addresses an important question of integrative biology.

Functional processes, such as the supply of oxygen and substrates to cells, are regulated to meet the actual demands of the cells primarily by varying the rates at which enzymes or organs, such as the heart, operate. But these rates very often reach an upper limit: $V_{max}$ for enzymes, and maximal heart frequency are two examples. These maximal rates determine the *functional capacity*, together with the total quantity of functional units that operate the system: the number of enzyme units or the volume of the heart chambers, to remain with the two examples. If maximal rates are "fixed," this implies that the overall capacity is, to an important extent, determined by *structural design* properties, and this then means that an adjustment of the functional capacity to altered needs requires morphogenetic processes.

It is now important to note that functional and structural regulation operate with very different time constants. Functional processes are regulated to acute changes in needs in time frames of milliseconds to minutes; structural adaptations to chronic changes in needs typically take at least days to months to be achieved. They also entail different costs: the cost of functional regulation is essentially limited to the incremental energy required to operate the system at higher levels; to make more structure requires not only energy to fuel the morphogenetic processes but also building materials, and space.

The most productive application of the hypothesis of symmorphosis is in the study of integrated functional systems where several organs are joined to serve one specific function. This can be a linear sequence, such as in the supply of oxygen from the lung to the mitochondria of muscle cells; or it can take the form of a branched network, as in the system for substrate supply for oxidative metabolism.

### Symmorphosis in the pathway for oxygen

The first case where we tested the hypothesis was the pathway for oxygen, which we considered a good test case for symmorphosis for three main reasons: (1) it serves one dominant vital function; (2) it involves a sequence of linked structures and the effect of design parameters on functional capacity can be defined (Figure 1.2); (3) the overall function has a measurable upper limit, $\dot{V}_{O_2 \, max}$, which varies between individuals and species so that the hypothesis can be tested by a comparative approach. This is further aided by the fact that variations in $\dot{V}_{O_2 \, max}$

result from three different effects or historical constraints: variations in body size cause body mass specific $\dot{V}_{O_2\,max}$ to increase by sixfold from cows to mice (allometric variation), whereas adaptation to different life styles result in 2.5-fold differences between horses and cows or dogs and goats (adaptive variation); finally, exercise training can increase $\dot{V}_{O_2\,max}$ by about 1.5-fold (induced variation).

These studies attempted to assess the quantitative relations between all parameters listed in Figure 1.2 by studying animals of varying body size and athletic versus sedentary species. The most interesting general result was that the structural parameters were rather precisely adjusted to overall functional capacity at all levels of the system, except for the lung (Figure 1.3), but that this was not achieved in a simple manner (Weibel, Taylor and Hoppeler 1991, 1992). In a broad comparative study from shrews to cows, we found that the volume of mitochondria increased with the same allometric slope as $\dot{V}_{O_2\,max}$, and in comparing athletic with sedentary species the mitochondrial volume was found to be proportional to the differences in $\dot{V}_{O_2\,max}$ (Figure 1.3a). As a result, it was concluded that mitochondria achieve the same rate of oxidative phosphorylation in all mammals, from shrews and mice to cows and horses. This was a simple case: at most levels above the mitochondria (Figure 1.2) we found that several parameters were adjusted part of the way. One example is the design of the muscle capillaries whose volume is adjusted to the muscle cells' need for oxygen only if the variability of the hematocrit, the $O_2$ carrying capacity of the blood, is taken into account. What is adjusted to the $O_2$ needs is the volume of capillary erythrocytes; this is achieved by increasing the concentration of erythrocytes *and* by making more capillaries. It appears indeed as eminently economic to split the effort of adaptation to higher needs between the two structures participating in this functional step: blood and vessels. Note, however, that had we looked only at single variables the full adjustment of design to functional demand would not have become apparent. There was one exception to this: the diffusing capacity of the lung was only partly adjusted to $\dot{V}_{O_2\,max}$ both in allometric and adaptive variation (Figure 1.3d) so that some of the species had some excess diffusing capacity, whereas others, mainly athletic and very small species, did not. The general conclusion was that the predictions of symmorphosis were supported for the internal compartments of the body whereas in the lung, the organ of interface to the environment, the hypothesis was not supported unconditionally.

## Symmorphosis in complex pathways

The case of the pathway for oxygen is comparatively simple: it constitutes a linear chain of structures connecting the source of $O_2$ with its target. The situation becomes more difficult when the hypothesis of symmorphosis is used to understand the design principles that govern the performance of complex pathways that have the basic structure of networks rather than a chain. One such example is the system for the supply of fuels for oxidative phosphorylation in muscle cells, which can take different paths by using different substrates from different sources and where some design components, such as the circulation of blood, serve several functions in parallel, one of them being the supply of $O_2$ for the combustion of these fuels. But, even in this instance, to structure the investigation into the effect of design on functional performance by submitting the hypothesis of symmorphosis turned out to be very productive as it allowed an assessment of the role of specific design properties in the overall perspective of systemic function (Taylor *et al.* 1996). The general conclusion reached was that the pathways for $O_2$ and fuels were designed to comply to different constraints and that this was compatible with the hypothesis of symmorphosis applied to network structures. Some of this will be discussed in Chapters 9.3 and 9.4 of this volume.

## Preparing for the debate

As rational as they seem, the theories of optimal design, and of symmorphosis in particular, have raised much criticism, mainly on the part of evolutionary biologists. It is contested, for example, whether evolution by natural selection can lead to optimal, rather than merely adequate, design. This controversy may be the result of different philosophies prevailing in evolutionary biology and in physiology – perhaps the different interests and the different emphasis in what are called ultimate and proximate causations of design. Can or should one of these be pursued at the exclusion of the other, or are they rather complementary views of nature? These are the fundamental questions we "ultimately" wished to address in order to clarify some of the controversial issues about optimization theory and its validity or heuristic usefulness, bearing in mind the fact that it holds in some instances but breaks down in others.

For this purpose we will first consider the criteria of optimal design under different points of view, and then discuss the processes and conditions of the evolution of species and how these affect the design of

*E. R. Weibel*

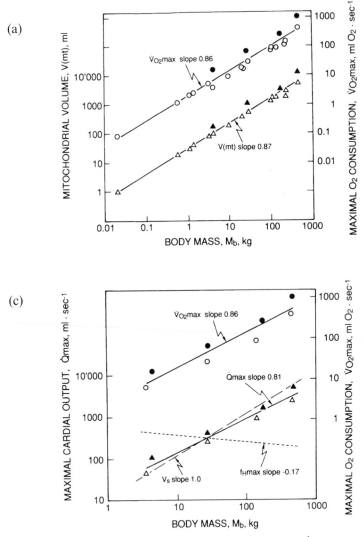

Figure 1.3. Four allometric plots showing the relation to $\dot{V}_{O_2\,max}$ of (a) total mitochondrial volume, (b) capillary volume, (c) cardiac output, and (d) pulmonary diffusing capacity. The open symbols represent sedentary species, and the solid symbols represent athletic species.

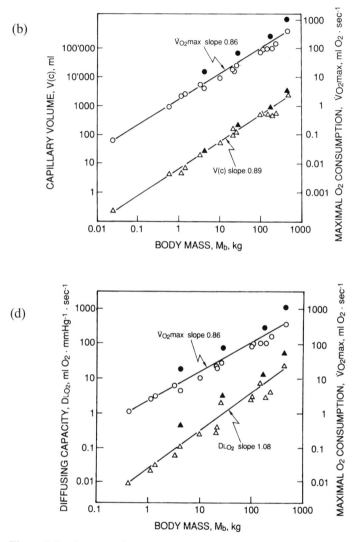

Figure 1.3.  (*continued*)

organisms. This will be followed by a number of case studies at the levels of cells, tissues, and organs, to find instances where optimization of design appears to prevail and where not. A particularly important part will be the discussion of design in complex functional systems for which the hypothesis of symmorphosis has been conceived. The examples used in this discussion range from enzyme systems and muscle cells, and the intestine and lung of different species, to the locomotor and nervous system, and the systems for energy supply.

## Further reading

Diamond, J. M. (1993) Evolutionary physiology. In *The Logic of Life: The Challenge of Integrative Physiology*, pp. 89–112. Eds. C. A. R. Boyd and D. Noble. Oxford University Press, Oxford, UK.

Dudley, R. and Gans, C. (1991) A critique of symmorphosis and optimality models in physiology. *Physiological Zoology*, **64**, 627–37.

Garland, T. and Huey, R. B. (1987) Testing symmorphosis: does structure match functional requirements? *Evolution*, **41**, 1404–9.

Lindstedt, S. L. and Jones, J. H. (1987) Symmorphosis and the concept of optimal design. In *New Directions in Ecological Physiology*, pp. 289–304. Eds. M. E. Feder, A. F. Bennett, W. W. Burggren and R. B. Huey. Cambridge University Press, New York.

Maynard Smith, J. (1978) Optimization theory in evolution. *Ann. Rev. Ecol. Syst.*, **9**, 31–56.

Taylor, C. R. and Weibel, E. R. (1981) Design of the mammalian respiratory system. *Respir. Physiol.*, **44**, 1–164.

Taylor, C. R., Weibel, E. R., Weber, J.-M., Vock, R., Hoppeler, H., Roberts, T. J. and Brichon, G. (1996) Design of the oxygen and substrate pathways. *J. Exp. Biol.*, **199**, 1643–1709.

Weibel, E. R. (Ed.) (1984) *The Pathway for Oxygen*. Harvard University Press, Cambridge, MA.

Weibel, E. R., Taylor, C. R. and Hoppeler, H. (1991) The concept of symmorphosis: a testable hypothesis of structure–function relationship. *Proceedings of the National Academy of Sciences*, **88**, 10357–61.

Weibel, E. R., Taylor, C. R. and Hoppeler, H. (1992) Variations in function and design: testing symmorphosis in the respiratory system. *Respir. Physiol.*, **87**, 325–48.

# 2

# How much structure is enough?

## 2.1 Overview

KNUT SCHMIDT-NIELSEN

The concept of symmorphosis suggests that structures should be matched to functional requirements. Consider, for example, the transport of oxygen from lungs to our muscles. The flow will be limited by whichever component of the transport system has the lowest carrying capacity. Any greater capacity in another component will be useless. Such never-used excess is costly in terms of materials as well as maintenance; it therefore seems rational that all components should have capacities matched to each other. This is what we mean by symmorphosis.

One difficulty in evaluating the concept of symmorphosis is the probability that any one component in a series may have multiple functions. Indeed, most systems are multifunctional. Thus, when we observe components that appear to negate the principle of symmorphosis, there may be some other dominant function. It is therefore desirable to establish the dominant function of any given component; for this objective the phenomenon of convergence can be helpful.

Convergence is the appearance of the same structural or functional feature in two or more different lineages, where the most recent ancestor common to any pair of those lineages did not have that feature. It is a remarkably common biological phenomenon. When a given function has evolved similarly but independently in multiple circumstances, it points to the likelihood that this particular function is dominant or essential.

Next, let us consider the concept of safety factors. Many structures, such as limb bones, have a strength in excess of the expected load. If the strength of a bone is twice as high as the maximum expected load, we say that the bone has a safety factor of two.

Are high safety factors desirable? Excess biological capacities would matter little if they cost nothing, but resources and space are limited and excess devoted to one component comes at the expense of another. For economy, it becomes a question of having enough but not too much.

Physiological safety factors have been little studied, but we shall see that the capacities of mechanisms that function serially tend to have similar safety factors. Furthermore, an upregulation of the capacities induced by increased loads results in similar increases in the constituent components.

However, the suggestion that all systems in a series should have equal capacities is not necessarily correct. We usually say that all links in a chain ideally should have equal strength but, surprisingly, this is not always true. Considering the inherent variability in any system (in this case, the strength of each link), we must also consider the cost of building and of maintaining a given structure. We then find that it may be advantageous to design the series with links of unequal strengths or capacities.

These conditions apply to systems for which a chain with a linear array of links is a suitable model. However, many functional systems are more complex, functional pathways may be branched, and some links may have several functions. To begin the discussion of symmorphosis it will nevertheless be helpful to consider simple systems in order to work out basic principles of how design relates to functional load. The first section asks how the function of structural features can be clarified by studying convergence, the second section introduces the notion of safety factors, and the third section introduces quantitative considerations of mechanical support systems.

## 2.2 Convergence as an analytical tool in evaluating design

### STEVEN VOGEL

The default hypothesis in physiology, generally used but rarely admitted, is the presumption that structures are ordinarily associated with specific functions. But the design of a structure may reflect anything from a single functional demand to a complex compromise among a diverse set of requirements. Comparative studies, especially ones directed toward patterns of evolutionary convergence, provide a way to determine the relative importance of different functions to structural design. The basic arguments run as follows:

(1) Natural selection acts more directly on organisms than on their microscopic and molecular constituents. Thus an assumption of efficient design is more useful in studies close to the organismic than to any other level of organization.

(2) This assumption is rarely made explicit because acceptance of the idea of evolution by natural selection discourages thinking in terms of goals and purposes.

(3) Evolutionary biology, in which phylogenetic analyses play an important role, usually finds convergence something to be detected in order to be circumvented.

(4) Convergence (that is, the independent development of similar design features) should play a more positive role in studies of function. Specifically, convergence provides a way to tell features that have important functional significance from features that do not. And the common aspects of the habit and habitat of the organisms involved point to the nature of that functional significance.

(5) Productive use of the concept of symmorphosis ordinarily requires a clear distinction between functions that strongly determine structural form and functions that, however crucial to the life of the organism,

13

are less important in its design. That distinction may be accomplished by careful examination of convergences among the various
components of a system whose symmorphosis is at issue.

That natural selection is at its most potent at or near the level of the
organism is too often obscured by arguments about its genocentric character. Most of the subcellular features common among organisms are
clearly ancestral rather than convergent. Conversely, cases of convergence at the organismic level are ubiquitous and involve both phylogen-
etically close and fairly distant organisms. Particularly dramatic cases are
the many strikingly similar marsupial and placental mammals, and the
leafless, fleshy stemmed, spinose, xeromorphic plants of both euphorbs
and cacti.

The physiologist does need reliable phylogenetic information in order
to make convergence a practical tool. Similarity of structure between
organisms is not enough: it is as important to know that the most recent
common ancestor lacked that similarity. Phylogenetic systematics is
based on shared, derived characters; inferring function from structural
similarity requires shared, non-derived, and thus convergent, characters.
Fortunately, the reliability of phylogenetic information has greatly
improved in recent years.

### Twist-to-bend ratios of woody structures

A simple case will illustrate how convergence can give an indication of
functional significance. Consider the problem of determining whether one
of a large number of mechanical variables is important in the mechanical
design of a group of load-bearing structures.

The "twistiness-to-bendiness ratio" of an elongate structure gives its
relative flexibility when faced with twisting (torsional) or bending
(flexural) loads. For a tree trunk, unless totally symmetrical and in a
horizontally uniform wind, a high ratio should be useful. Minimizing
bending will keep the trunk upright and the center of gravity in place,
while permitting twisting may actually reduce exposure to bending stress.
Values for trunks are about five times those for ordinary, isotropic materials (Table 2.1). But is this situation a feature on which fitness depends
or an indirect consequence of the anisotropic character of wood? Finding
a high ratio in a diversity of trees strengthens the argument for functional
significance only slightly. Better is a clear convergence onto high ratios of
the trunks of two lineages of trees and the culms of bamboo, each inde-

Table 2.1. *Values of the twistiness-to-bendiness ratio (flex-ural to torsional stiffness, EI/GJ)*

| | |
|---|---|
| Aluminum circular cylinder | 1.3 |
| Tree trunks, angiosperms | 8.7 |
| Tree trunks, gymnosperms | 5.3 |
| Culms, bamboo | 8.6 |
| Grooved petioles | 5.0 |
| Sparrow wing feathers | 4.8 |
| Locust tibias | 6.4 |
| Grapevines | 2.7 |
| Tree roots | 2.3 |

*Source*: Vogel (1992, 1995 and unpublished material).

pendently evolved from herbaceous forms. Still better evidence for significance comes from cases where the functional imperative does not hold: vines and roots, woody structures for which ease of twisting relative to bending should not matter, have much lower ratios.

Additional evidence that the variable is important comes from still broader convergences. Petioles, especially short ones, ought to twist easily so leaves can cluster in winds; but they should resist bending to keep leaves from drooping. The wing feathers of a bird must twist oppositely on up- and downstrokes, but they can only support the bird if resistant to bending. The hind tibias of a locust must resist bending if it is to jump, but they ought to twist easily to minimize torsional loads on the twist-sensitive joints. All these likewise have high twist-to-bend ratios (Table 2.1). Even more interesting, all of these latter achieve their high ratios in part by geometric adjustments – lengthwise grooves and so forth – rather than by the purely material adjustments of uniformly cylindrical tree trunks. What matters is the value of the variable, not how it is achieved.

## Murray's law and vascular systems in plants

In the previous example only one factor was examined. Convergence, though, can point to functional significance and probable symmorphosis where several factors are involved. A large number of systems in which liquid flows laminarly through branching arrays of cylindrical pipes have converged on the same rule for relative pipe size. The rule (usually called "Murray's law") is simple: the sum of the cubes of the vessel radii is the

same at any level of the branching array. The rule simultaneously minimizes the cost of flow, obtained from the Hagen–Poiseuille equation, and
a cost, proportional to volume, that probably reflects material investment, construction, and maintenance. Murray's law works well (Figure
2.1) for the circulatory systems of various vertebrates, cephalopods, and
crustaceans; for the aquiferous system of sponges; and for the astrorhizae
of fossil stromatoporoids (LaBarbera 1990, 1995).

Neither common origin nor genetic or developmental constraints are
likely. First, the arrangement for maintaining fidelity to Murray's law in
vertebrate blood vessels, endothelial cell proliferation, is not available to
most of the others. Second, the rule is less precisely followed by systems
in which branching is especially frequent and the Hagen–Poiseuille equation is less strictly applicable, such as the coronary circulations in mammals and cephalopods. Third, in a system that uses diffusion rather than
bulk flow, the tracheae of some larval moths, or airways in the acinus of
the lung, total cross-sectional area is conserved instead. Thus the highly
convergent design is good evidence that the cost of liquid transport matters to the fitness of these organisms.

Further convergence tells us still more. The vessels of xylem in the
leaves of sunflowers follow Murray's law (Canny 1993). More significant
than mere addition of another kingdom is what the case implies about the
two components, flow and volume, of the economic argument. Sap is
mainly drawn up tall vascular plants by evaporation within the leaves,
a process that uses solar energy without metabolic intervention. So the
cost of flow should not matter very much, if at all; in any case, flow and
gravitational work together consume less than a thousandth as much
energy as evaporation. About the only relevant variables left are the
weight of the sap and the volume of the walls of the vessels. Thus this
case provides evidence that the volume component alone can be sufficient
to compel adherence to Murray's law.

Murray's law, though, does not describe, in trees, the sizes of the long-
distance vessels of trunks and stems. The latter have clearly converged,
but on a different rule. Pressure drops from flow at normal rates are
consistently about one atmosphere per ten meters of height, or 10,000
$Pa\,m^{-1}$ for an order of magnitude range of vessel diameters
(Zimmermann 1983). What appears to determine this peculiar constancy
is the gravitational pressure drop of the same 10,000 $Pa\,m^{-1}$. Since the
latter is unavoidable, reducing the pressure drop due to flow much below
it would give a much diminishing fractional reduction in their sum. In
addition, the wider vessels needed to decrease the pressure drop due to

(a)

(b)

(c)

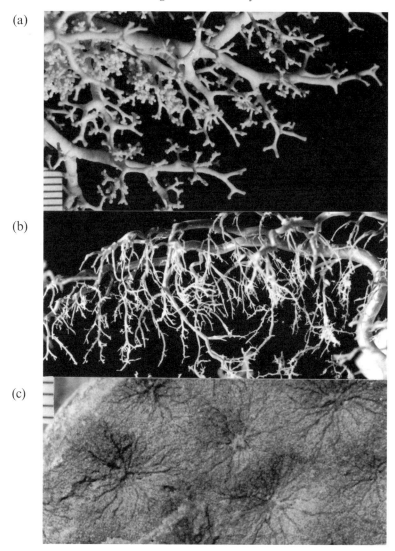

Figure 2.1.    Three arrays of branching vessels: (a) the small airways of a human lung; (b) coronary arteries of human heart; (c) the astrorhizae imprinted on the inner surface of fossil stromatoporoids, an invertebrate group of uncertain affinities. (Figure 2.1(c) courtesy M. LaBarbera.)

flow would be more susceptible to cavitational embolism. In short, the overall system appears to show symmorphosis, but the convergences show that different rules predominate in different places.

## The shapes of broad leaves

In each previous case a single function dominated design. But structures with several major functions are common and present a special problem for application of symmorphosis. In particular, trouble arises where the fitness of the organism is so dependent on more than one function of a structure that each function seriously constrains its design. But careful attention to convergence can reveal how much each function bears on the overall form of the structure – or which functional systems are most likely to show symmorphosis.

Consider the shapes of the broad leaves of ordinary, deciduous trees. Leaves come in a broad range of shapes and sizes, and they must accomplish several important tasks that make separate demands on their form. They must intercept light efficiently, work as cantilever beams in still air and light winds, withstand substantial intermittent wind without being damaged or transmitting excessive force to branches and trunk, provide channels for distribution and collection of liquid, avoid excessive heating if exposed to full sunlight and nearly still air, and so forth.

One may ask what features characterize leaves that can tolerate direct sunlight without wilting downward or otherwise avoiding a horizontal orientation in which illumination is maximal and convective heat dissipation least effective. Such leaves are small, narrow, elaborately scalloped, or divided into small leaflets. The pattern confirms the indication from measurements of temperature on both models and real leaves that overheating is a significant hazard for leaves in which heat loss is predominantly convective (Vogel 1970). Note that examination of patterns of convergence and measurements under controlled conditions are complementary. In this case, the convergence might reflect some other variable entirely, whereas the measurements might just see solutions so effective that they obscure the underlying problem.

One may alternatively ask how leaf shape and structure minimize drag in occasional high winds. Leaves with relatively long petioles and blades that extend proximally (adaxially) as lobes on either side of the petiole's attachment point (leaves of generally cardioid shape) reconfigure into aerodynamically stable cones of low drag (Vogel 1989). That convergent correlation of petiole length and blade shape (the latter occurring in at least fifteen families; cf. Table 2.2) thus takes on functional significance. Pinnately compound leaves achieve even lower drag in high winds as the individual leaflets reconfigure into stable whole-leaf cylinders. Pinnate compoundness is, as noted earlier, a form that also does well at convec-

Table 2.2.  *Families that include but do not entirely consist of trees whose leaves have both substantial petioles and leaf blades that extend proximally from the petiolar attachment, with an example from each family*

| Family | Tree | Common name |
|---|---|---|
| Aceraceae | *Acer rubrum* | Red maple |
| Betulaceae | *Betula populifolia* | Gray birch |
| Bignoniaceac | *Catalpa speciosa* | Northern catalpa |
| Caricaceae | *Carica papaya* | Papaya |
| Hamamelidacreae | *Liquidambar styraciflua* | Sweetgum |
| Leguminoseae | *Cercis canadensis* | Eastern redbud |
| Magnoliaccae | *Liriodendron tulipifera* | Yellow poplar |
| Malvaceae | *Hibiscus tiliaceus* | Hibiscus |
| Moraceae | *Morus rubra* | Red mulberry |
| Platanaceae | *Platanus occidentalis* | Sycamore |
| Rosaceae | *Crataegus pruinosa* | Frosted hawthorn |
| Salicaceae | *Populus deltoides* | Eastern cottonwood |
| Stericuliaceae | *Firmiana platanifolia* | Chinese parasol tree |
| Tiliaceae | *Tilia americana* | American basswood |
| Ulmaccae | *Celtis tenuifolia* | Georgia hackberry |

tive heat dissipation, perhaps explaining its prevalence among the canopy trees of tropical rain forests. By contrast, the deeply scalloped leaves of some large oaks, while very good for heat dissipation in still air, have high drag and are unstable in high winds. This last result suggests searching for other advantages that might be associated with their shape. Again, examining convergences and measuring the relevant functional variables are complementary, with the results of one often suggesting investigations using the approach of the other.

## Conclusions

The concept of symmorphosis represents a contribution of comparative physiology to evolutionary studies; that it has been controversial is, in part, just a result of entry into what has always been a more disputatious arena. The concept of convergence is the opposite, a tool that evolutionary biology provides to comparative physiology.

Comparative approaches to functional problems have long been used to detect the invariants underlying diversity and complexity, and to infer functional principles from regular variation among organisms of different habits and habitats. In blissful ignorance of evolution and phylogenies but to great effect, William Harvey (1628) looked at reptiles, amphibians, fish, worms, insects, crustaceans, and mollusks as he attempted to see quantitative rationality in circulatory systems. As Harvey put it, "If only anatomists were as familiar with the dissection of lower animals as with that of the human body, all these perplexing difficulties would, in my opinion, be cleared up."

Armed now with reliable phylogenetic information, the physiologist can do much better. Organisms can be selected with similarities that are clearly not derived from common ancestry and that are almost certainly driven by functional imperatives. Where structures have multiple functions, such convergences should prove especially useful in determining which functional demands are most critical in determining structure and thus which structural features should show symmorphosis.

### Further reading

Canny, M. J. (1993) The transpiration stream in the leaf apoplast: water and solutes. *Phil. Trans. Roy. Soc. Lond. B*, **341**, 87–100.

Harvey, W. (1628) *Anatomical Studies on the Motion of the Heart and Blood in Animals*. Translated by C. D. Leake, 1941. C. C. Thomas, Springfield, IL.

LaBarbera, M. (1990) Principles of design of fluid transport systems in zoology. *Science*, **249**, 992–1000.

LaBarbera, M. (1995) The design of fluid transport systems: a comparative perspective. In *Flow-Dependent Regulation of Vascular Function*, pp. 3–27. Eds. J. A. Bevan, G. Kaley and G. M. Rubany. Oxford University Press, New York.

Vogel, S. (1970) Convective cooling at low airspeeds and the shapes of broad leaves. *J. Exp. Bot.*, **21**, 91–101.

Vogel, S. (1989) Drag and reconfiguration of broad leaves in high winds. *J. Exp. Bot.*, **40**, 941–8.

Vogel, S. (1992) Twist-to-bend ratios and cross-sectional shapes of petioles and stems. *J. Exp. Bot.*, **43**, 1527–32.

Vogel, S. (1995) Twist-to-bend ratios of woody structure. *J. Exp. Bot.*, **46**, 981–5.

Zimmerman, M. H. (Ed.) (1983) *Xylem Structure and the Ascent of Sap*. Springer-Verlag, New York.

## 2.3  Evolution of biological safety factors: a cost/benefit analysis

### JARED M. DIAMOND

A central problem of integrative biology is to understand quantitative biological design at every level, from the molecular level to the whole animal. The elements of quantitative design are familiar. For example, molecular biologists measure activities of enzymes and transporters; cell biologists measure numbers of ion permeation channels or receptors per square micrometer of membrane; anatomists measure masses of organs and surfaces of tissues; and physiologists measure functions such as the diffusing capacity of the lung.

How are all of these quantities *ultimately* determined by the loads imposed upon them? For example, how did we humans come to be genetically programmed such that the glucose transporter makes up about 0.1 percent of our intestinal brush-border protein, when it could instead have made up 1 percent as in hummingbirds, or 0.001 percent as in snakes? That quantity of glucose transporter in each species is evidently a matter of some consequence, since its interindividual variation in each species is only about 15 percent. Would there be serious penalties for possessing either double or one-half of the usual amount of glucose transporter? What are those penalties? How has natural selection fine-tuned our glucose transporter capacity so as to match it to the glucose loads that our particular lifestyle places on it? Are we designed with capacities in excess of loads, and if so, by how much?

### Safety factors in engineering and biological design

Let us consider the corresponding problem for engineered structures. Engineers define the "safety factor" as the ratio of a component's capacity (or strength, or performance) to the maximum expected load upon that component during operation. A typical example would be the ratio

21

Table 2.3.   *Safety   factors of some engineered structures*

| | |
|---|---|
| Cable of fast passenger elevator | 11.9 |
| Cable of slow passenger elevator | 7.6 |
| Cable of slow freight elevator | 6.7 |
| Wooden building | 6 |
| Cable of powered dumbwaiter | 4.8 |
| Steel building or bridge | 2 |

of the load under which a structure (for example, an elevator cable) would break to the expected maximum load. Table 2.3 summarizes safety factors for some actual engineered structures; most of their safety factors fall within the range 2 to 12.

The proximate explanation for this variation in safety factors is clear: it involves the strength selected for materials, the size selected for components, and the overall design selected for the whole object. The ultimate explanation for the variation is also clear: engineers specify safety factors after doing a cost/benefit analysis.

One consideration is the penalty for failure: the higher the penalty, the higher the specified safety factor. For example, the penalty is much higher for a cable snapping on a passenger elevator than on a freight elevator, so engineers specify a higher safety factor for passenger elevators. Other considerations are variation in capacity (for example, due to variations in material strength), the variation in expected load, and the amount of deterioration expected in a component that cannot be readily replaced or repaired (such as the steel beams inside a skyscraper). The safety factor specified increases with each of these quantities. On the other hand, the higher the cost of materials, and the more that the costs of maintenance and of use increase with component strength, the lower will be the safety factor specified.

In the case of these engineered safety factors, the arena of the decision is the marketplace with its competing manufacturers, and the selective factor is consumer choice. The biological analog is the marketplace of competing individuals and species, and the selective factor is natural selection operating through differential survival and reproduction. The result is some selected ratio of capacity to load for each biological component.

Of course, excess biological capacity would be desirable if it cost nothing. But the resources available to an animal are generally finite, so that excess devoted to one component comes at the expense of another component. That is, there is an evolutionary basis to biological capacities, but the magnitudes of biological safety factors have been little explored. Do biological capacities typically exceed their loads by 10, 100, or 1,000 percent? How much biological machinery constitutes "enough but not too much"? These questions define a research program consisting of three subprograms: (1) measuring safety factors for biological components; (2) explaining the observed variation in safety factors; and (3) explaining the costs leading to evolutionary elimination of excessive safety factors.

Safety factors are most straightforward to define and measure for load-bearing biological structures such as bones, tendons, shells, and spiders' webs, which are directly analogous to engineered load-bearing structures (Table 2.4). In these the ratio of breaking strength to actual operating stress is the safety factor, and most fall within the range 1.3 to 6.

### Physiological safety factors

An example of a physiological safety factor is that for the renal glucose transporter. The renal transport maximum (so-called $T_m$) for glucose is approximately twice the normal renal filtration rate of glucose, yielding a safety factor of approximately 2.

Another example is the value of approximately 2 for milk production (or teat number, its anatomical surrogate). Aristotle noted over 2,000 years ago that teat number among mammal species varies from 2 to 16, but that in each species the teat number is approximately twice the normal litter size and equal to the maximum natural litter size, giving a safety factor of 2 for normal operation.

In order to obtain physiologically meaningful values of safety factors at the molecular level, my colleagues and I have measured safety factors for hydrolases and transport proteins of the small intestinal brush-border. By the everted sleeve method one can measure the functional capacity of the brush-border as $V_{max}$ values for hydrolases (such as sucrase) and transporters (such as sugar and amino acid transporters) in a preparation with epithelial and brush-border structure intact. We then measure the loads upon these molecules as average rates of dietary nutrient intake. From this we then calculate the molecular safety factors as ratios of $V_{max}$ to intake, while experimentally manipulating both the capacities, by surgical partial resection of the intestine, and the loads by maintaining animals under conditions associated with high-energy output (low ambient temperature, lactation, or exercise), thus stimulating the animal to increase its food intake by up to five times normal values.

Table 2.4.  *Safety   factors of some biological structures*

| | |
|---|---|
| Human pancreas | 10 |
| Cat intestinal brush-border arginine transporter | 7 |
| Wing bones of a flying goose | 6 |
| Leg bones of a galloping horse | 4.8 |
| Human kidneys | 4 |
| Leg bones of a hopping kangaroo | 3 |
| Mouse intestinal brush-border glucose transporter | 2.7 |
| Mouse intestinal brush-border sucrase | 2.7 |
| Leg bones of a running ostrich | 2.5 |
| Human small intestine | 2 |
| Lungs of a lazy big cow | 2 |
| Breasts of most mammals | 2 |
| Drag line of a spider | 1.5 |
| Backbones of a weightlifter lifting weights | 1.0–1.7 |
| Shell of squid | 1.3–1.4 |
| Lungs of a fast small dog | 1.25 |

Figure 2.2 shows the results of an experiment where dietary nutrient intake by lactating mice was increased by increasing the number of pups they nurse and by lowering the ambient temperature. The safety factor for glucose intake declined as a greater fraction of uptake capacity was used to accommodate the greater nutrient load.

We have measured safety factors for various transporters and hydrolytic enzymes of the intestinal brush-border, in numerous vertebrate species (mice, rats, rabbits, cats, chickens, frogs, and fish), at different ages from birth or hatching to adulthood, and under different energy loads. The measured safety factors prove typically to be around 2 and range up to 7.

### The meaning of biological safety factors

My discussion has so far been simplistic, in order to make the basic concept clear. Let us now consider some of the complications that may occur. Consider first the problems posed by a system of many components arranged in series. An evolutionary argument might suggest that the capacities of the components would be matched to each other, and that they would have evolved to possess approximately equal safety factors. If one component has a much lower capacity than other components, this component would be rate limiting. The excess quantities of all other components would be largely wasted, because of the bottleneck

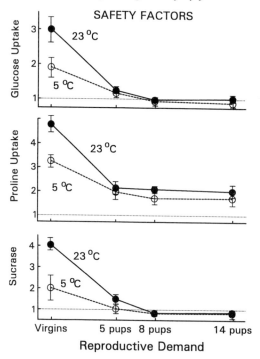

Figure 2.2. Safety factors for glucose uptake, proline uptake, and sucrase in virgins and lactating mice at 23°C (●) and 5°C (○) as a function of reproductive demand. (Reproduced from Hammond *et al.* (1994) *Physiol. Zool.*, **67**, 1479–506. © by The University of Chicago.)

imposed by the limiting component. Are the transporter capacities rate-limiting and the hydrolases present in excess, or vice versa, or do the transporters and hydrolases possess similar capacities? It turns out that both proteins possess similar $V_{max}$ values, corresponding to safety factors around 2.7. Thus, neither constitutes a single rate-limiting step.

This conclusion is supported by the fact that, with increasing dietary carbohydrate levels, there is upregulation of the quantities or $V_{max}$ values of at least four steps in series in carbohydrate digestion: pancreatic amylase, brush-border sucrase and maltase, the brush-border glucose and fructose transporters, and basolateral glucose transporter. Some biological systems do regulate their capacities depending on their loads. These are the phenomena of load-dependent tissue growth and atrophy, and enzyme and transporter induction and repression. For example, the mass of the small intestine (hence the capacities of all of its transporters and enzymes) increases with conditions stimulating increased food intake, such as lactation or low ambient temperatures. The quantitative details of this capacity/load relation for the mouse intestinal brush-border glucose transporter are such that the transporter's safety factor declines from about 2.7 at low loads to near 1.0 with increasing loads, and then remains at 1.0 up to the highest loads that we have been able to produce experimentally.

## Cost/benefit considerations

Tissue atrophy and enzyme repression under low loads suggest that excess tissues and capacities would incur significant costs. What are the relevant costs? That is, what are the selective factors leading to elimination of excessive biological capacities? This is one of the major unsolved problems of biological design.

Biologists usually think of costs as being measured in the currency of biosynthetic energy for synthesis or maintenance. That reasoning is plausible in the case of components contributing a significant percentage of an animal's entire energy budget (for example, kidney, liver, and intestine), but such energy-based reasoning can hardly apply to all of the very minor components of our bodies. For example, the small intestinal brush-border glucose transporter constitutes only 0.1 percent of the protein of the brush-border of the intestine, and probably much less than 0.0001 percent of our total energy budget. It would make no difference to the whole animal's energy budget if the amount of this quantitatively utterly insignificant component were doubled. Why is the glucose transporter nevertheless regulated with an interindividual coefficient of variation of only about 15 percent?

A suggested alternative involves considerations of space, which is also in short supply. The insides of an animal are crammed full of organs, so that the volume of one organ could not increase without either decreasing the volume of other organs or increasing the size of the whole animal. These standing-room-only conditions apply at the molecular level as well. For example, while the glucose transporter contributes a trivial percentage of the whole animal's energy budget, it is one of the major membrane-spanning proteins of the brush-border. One cannot increase a membrane's proportion of one membrane-spanning transport protein, without decreasing the proportion either of other transport proteins or of lipid bilayer, either of which would affect membrane function. Similarly, most of the surface of the intestinal brush-border is covered with hydrolases that include sucrase, maltase, lactase, and peptidases. Since these hydrolases are crammed together, one cannot increase the quantity of one without reducing the quantity of others.

Thus, space considerations may create severe penalties for excessive unutilized capacities of even quantitatively very minor components. Hence I would suggest that, while considerations of biosynthetic energy may explain the modest safety factors of large expensive components, considerations of space are what limit the safety factors of minor components.

### Further reading

Buddington, R. K. and Diamond, J. M. (1989) Ontogenetic development of intestinal nutrient transporters. *Ann. Rev. Physiol.,* **51**, 601–9.

Diamond, J. M. (1993) Evolutionary physiology. In *Logic of Life: The Challenge of Integrative Physiology*, pp. 89–111. Eds. D. Noble and C. A. R. Boyd. Oxford University Press, Oxford.

Diamond, J. M. and Hammond, K. A. (1992) The matches, achieved by natural selection, between biological capacities and their natural loads. *Experientia,* **48**, 551–7.

Dykhuizen, D. (1978) Selection for tryptophan auxotrophs of *Escherichia coli* in glucose-limited chemostats as a test of the energy conservation hypothesis of evolution. *Evolution,* **32**, 125–50.

Ferraris, R. P. and Diamond, J. M. (1989) Substrate-dependent regulation of intestinal nutrient transporters. *Ann. Rev. Physiol.,* **51**, 125–41.

Hammond, K. A. and Diamond, J. M. (1992) An experimental test for a ceiling on sustained metabolic rate in lactating mice. *Physiol. Zool.,* **65**, 952–77.

Hammond, K. A., Konarzewski, M., Torres, R. and Diamond, J. M. (1994) Metabolic ceilings under a combination of peak energy demands. *Physiol. Zool.,* **67**, 1479–1506.

Toloza, E. M., Lam, M. and Diamond, J. M. (1991) Nutrient extraction by cold-exposed mice: a test of digestive safety margins. *Am. J. Physiol.,* **261**, G608–20.

Weibel, E. R., Taylor, C. R. and Hoppler, H. (1991) The concept of symmorphosis: a testable hypothesis of structure–function relationship. *Proc. Natl. Acad. Sci. USA,* **88**, 10357–61.

# 2.4 Symmorphosis and safety factors

## R. McNEILL ALEXANDER

A chain is as strong as its weakest link. Therefore, we may conclude, it is pointless to design one link to be stronger than another. It will be shown that this conclusion is sometimes false.

The chain is a convenient model for body parts that cooperate in a function. The bones of a leg function as a set, and if one breaks the leg it is as useless as a broken chain. The systems that transport oxygen to our muscles (lungs, breathing muscles, blood, heart, capillaries, and mitochondria) function together, and rates of oxygen uptake are limited by whichever one of them has the smallest oxygen-carrying capacity.

The principle of symmorphosis may lead us to expect every bone in a leg to be capable of withstanding the same force on the foot, and every link in the respiratory chain to be capable of transporting oxygen at the same rate, but we must consider the possibility that different links may be best designed with different safety factors.

A smith making a chain would incorporate a safety factor, making it somewhat stronger than he expected it to need to be. Nevertheless, it might break in use because human error and imperfect materials make the strength of a chain of given design somewhat unpredictable, and because the maximum load that will act on a chain is also somewhat unpredictable. Suppose that a particular chain had to be made partly of gold links and partly of iron ones. The obvious design would have the gold and iron links equally strong, but if the smith made the expensive gold links very slightly weaker he could afford to make the cheap iron ones considerably stronger, which might make the chain as a whole stronger, for the same price.

## Variability, cost and safety factors

This unexpected suggestion depends on the variability affecting individual links being to some extent independent. If the strengths of the links were precisely predictable and all the variability were in the load then, plainly, it would always be the link designed to be weakest that failed: the links should be designed to be equally strong. If, however, link strengths were unpredictable, it might be advantageous to design the iron links to be stronger.

The safety factor of a link is the ratio of the strength it is designed to have, to the maximum load expected to act on it. The probability $P_{link}$ that the link will break is a function of the safety factor $S$ and a variability parameter $\nu$ which describes the extent to which strength and load are unpredictable. By definition, the probability of breakage is 0.5 when the safety factor is 1. It is 1 when the safety factor is zero (a link with no strength will certainly break) and zero when the safety factor is infinite. We can model this relationship by writing

$$P_{link} = 1 - F[(1/\nu)\log_e S] \qquad (1)$$

where $F$ represents the cumulative normal distribution function. Figure 2.3 shows graphs of $P_{link}$ against $S$ for different values of $\nu$.

Suppose that our smith has sufficient funds to make both links of his chain with safety factor $\bar{S}$. If he chooses to make the gold link a little weaker, giving it safety factor $(\bar{S} - \Delta S)$, he can afford to make the iron

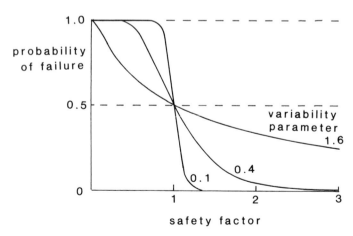

Figure 2.3. The probability that a link will fail as a function of its safety factor, for three different values of the variability parameter, $\nu$.

link stronger, with safety factor $(\bar{S} + c.\Delta s)$ where $c$ is the ratio of the prices (for the same strength) of gold and iron. The probability that the gold link will *not* fail is then $F[(1/\nu_{gold}) \log_e(\bar{S} - \Delta S)]$ and the probability that the iron link will not fail is $F[(1/\nu_{iron}) \log_e(\bar{S} + c.\Delta S)]$. If these probabilities are perfectly correlated (as will happen, for example, if all the variability is in the load which is applied to both links), the probability of failure for the chain is simply the probability of failure of the weaker link.

However, if the probabilities are wholly uncorrelated, the probability that the chain will break is

$$P_{chain} = 1 - F[(1/\nu_{gold}) \log_e(\bar{S} - \Delta S)]F[(1/\nu_{iron}) \log_e(\bar{S} + c.\Delta S)]. \quad (2)$$

(The probabilities will be uncorrelated and this equation will apply if the load is precisely predictable and all the variability is due to the unpredictable properties of the links. )

Equation (2) has been used to calculate probabilities of breakage for chains whose links have unequal safety factors. Figure 2.4(a) confirms that an advantage may be gained by making the gold link weaker and the iron one stronger. Remember that equation (2) depends on the assumption of uncorrelated probabilities of failure: the conclusion, that the gold link should be made weaker, would not apply if the probabilities were perfectly correlated.

Further calculations confirm that the greater the cost differential, the greater the difference in safety factor should be. Less obviously, the greater the variability parameters and the smaller the mean safety factor $\bar{S}$, the greater the ratio of safety factors should be. ($\bar{S}$ is the weighted mean – weighted according to the cost of the materials – of the safety factors of the two links.)

## Biological "chains"

The prediction, that the expensive links in a chain should be made weaker than the cheap ones, may explain a veterinary observation. In horse racing, leg bones may break as the result of a fall or while the horse is still running, prior to falling: the latter accidents probably result from bone fatigue. Falling may impose enormous forces on the bones, and whichever bone these forces act on will get broken. John Currey has suggested that fatigue fractures probably give a better indication of the relative safety factors of the bones. Fractures occurring prior to falling are more common in the cannon bones (the long bones of the feet) than

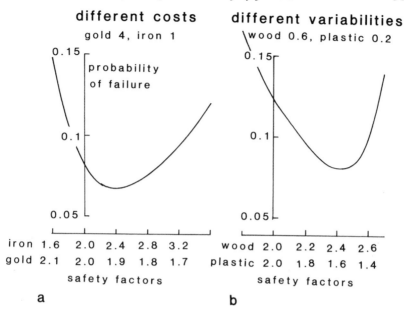

Figure 2.4. The probability that a mixed chain will fail plotted against the safety factors of its links, calculated from equation (2). It is assumed that if one link is made stronger, the other must be made sufficiently weaker to keep the total cost constant. In (a) the materials have different costs but equal variabilities; and in (b) they have equal costs and different variabilities. Gold is more expensive than iron and wood is more variable than plastic. The parameters used for the calculations are (a) $c = 4$, $\nu_{gold} = \nu_{iron} = 0.4$, $\bar{S} = 2$; (b) $c = 1$, $\nu_{plastic} = 0.2$, $\nu_{wood} = 0.6$, $\bar{S} = 2$.

in the bones higher up the legs (Figure 2.5). Thus it seems likely that the cannon bones are designed to lower safety factors than the other long bones.

All the bones are made of the same material, but the main cost of having a stronger bone is not the cost of the extra material: it is the energy cost of having to move additional mass when running. As the legs swing backwards and forward, their lower parts have to be accelerated and decelerated through much larger ranges of velocity than do their upper parts. Therefore, strength is more costly in the bones of the lower parts of the leg and we may expect the cannon bones to be built to lower safety factors than the bones higher up the leg.

This conclusion would not apply if the probabilities of failure of the bones were perfectly correlated. Correlation is imperfect because the stresses in a bone, due to a given ground force, depend on the angle of

*R. McN. Alexander*

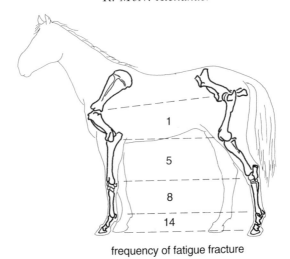

1

5

8

14

frequency of fatigue fracture

Figure 2.5. The frequency of fatigue fracture in a horse's leg bones increases towards the periphery. (Data from Currey 1984.)

the bone relative to the force; and the angles of the bones can vary independently. Also, the strength of each bone is to some extent unpredictable. The conclusion, that cannon bones should have relatively low safety factors, seems sound.

In the case of the oxygen-transport chain we are not concerned with the probability of breakage, but with the probable limit to the rate of oxygen transport. In an emergency, when the animal runs as fast as possible, its rate of oxygen uptake will rise until it reaches the limit for one of the links of the chain. Any further increase will be impossible.

Consider a two-link oxygen-transport chain. Resources are sufficient to give the links weighted mean oxygen-carrying capacity $\bar{C}$. Thus if an expensive (gold) link is designed to carry oxygen at a maximum rate $(\bar{C} - \Delta C)$, the cheaper (iron) link can be designed for a rate $(\bar{C} + c.\Delta C)$. Assume that both these rates are somewhat unpredictable, with variabilities represented by the parameters $\nu_{\text{gold}}$, $\nu_{\text{iron}}$. The probability that the maximum rate at which the chain can transport oxygen will be less than or equal to $R$ is then

$$P_{\text{chain}} = 1 - F\{(1/\nu_{\text{gold}}) \log_e[(\bar{C} - \Delta C)/R]\}F\{(1/\nu_{\text{iron}}) \log_e[(\bar{C} + c.\Delta C)/R]\}. \tag{3}$$

(This equation could be made identical to equation (2) by writing $\bar{C}/R = \bar{S}$ and $\Delta C/R = \Delta S$.)

The expected maximum rate of uptake of oxygen can be found by setting $P_{\text{chain}} = 0.5$ and solving equation (3) for $R$. This can be done for different values of $\Delta C$ (keeping $C$ constant) to find the design that gives the highest expected rate of oxygen transport. By doing this for a cost factor $c = 4$ and variabilities $\nu_{\text{gold}} = \nu_{\text{iron}} = 0.4$ (the same values as were used in Figure 2.4a) one finds that the iron link should be designed to have a carrying capacity 1.45 times that of the gold link. However, this optimum ratio of carrying capacities is highly dependent on the variabilities. If the capacities are more predictable, with $\nu_{\text{gold}} = \nu_{\text{iron}} = 0.1$, the iron link should be designed to have only 1.12 times the capacity of the gold one.

It seems likely that the links in the oxygen-transport chain differ in cost. The principal costs to be considered here are not the initial costs of forging the links but the metabolic costs of maintaining them. These may include costs incurred elsewhere in the body by producing materials that are transported for use to the tissues in question, but for lack of better information we may use the metabolic rates of tissues as indicators of their costs. The metabolic rate of the lungs of a mammal is much less than the resting metabolic rate of the entire skeletal musculature. Thus the oxygen-diffusion surface of the lungs is a relatively cheap link in the chain and we may expect it to be designed for a higher carrying capacity than the relatively expensive mitochondria.

Though the links in the oxygen-transport chain differ in cost, we cannot expect them to be designed for different carrying capacities unless these capacities are somewhat unpredictable. If its value is small, the variability $\nu$ of a lognormal function is approximately equal to the coefficient of variation (standard deviation divided by mean). Thus $\nu = 0.1$ (an example used in a calculation above) implies a coefficient of variation of 0.1. There seem to be no data to tell us whether the coefficients of variation of oxygen-carrying capacities are greater or less than that.

## Predictability and safety factors

So far, we have considered chains whose links differ in the costliness of the material, represented by gold and iron. Another possibility we should consider is that the materials may cost the same (for given strength) but that one may be less predictable than the other. They may, for example, be wood and plastic. The properties of wood are rather unpredictable because of variable orientation of the grain, because it may be weakened by knots, and so on; consequently, engineers build wooden structures with high factors of safety. In contrast, human-made plastics may be more predictable.

Figure 2.4(b) shows the results of calculations using equation (2) again, but with the cost factor $c$ equal to one (the two materials are equally costly) and with different values for the two variability parameters (one material is less predictable than the other). It shows that to minimize the probability of failure, the link made of the more variable (less predictable) material should be given a higher safety factor. Further calculations show that if the mean safety factor $\bar{S} = 1$, the safety factors should be kept equal but that for larger values of $\bar{S}$, the optimum ratio of safety factors increases as $\bar{S}$ increases. If the more variable material is also the less expensive, the effect is enhanced.

Let us return to the example of horse leg bones. We have seen that strengthening the cannon bone is more expensive than strengthening the bones of the upper leg, in that it adds more to the energy cost of running. It seems possible that, in addition, stresses may be less predictable for the cannon bone, because its orientation may be more variable. If so, this may partly counteract the tendency for the cannon bone to acquire a lower safety factor.

### Conclusions on symmorphosis

Symmorphosis implies that all the bones of a leg should be designed to withstand the same force on the foot, and that successive components in an oxygen-transport system should be designed to have the same carrying capacity. This section has shown that the conclusion is false unless the strengths and carrying capacities are precisely predictable. If they are subject to unpredictable variations more expensive ("gold") links in a chain should be designed to be a little weaker so that cheaper ("iron") ones can be made stronger. Also, more predictable ("plastic") links should be made weaker so that less predictable ("wood") ones can be made stronger. So far we have little evidence of these principles applying in nature – but, so far, little effort has been made to look for such evidence. Weibel and Taylor's classic verification of the principle of symmorphosis in oxygen-transport systems focused on the exponent $b$ in allometric equations of the form $y = ax^b$. To test the ideas presented in this section we must also consider the factor $a$.

## Further reading

Alexander, R. McN. (1981) Factors of safety in the structure of animals. *Science Progress, Oxford,* **67**, 109–30.

Alexander, R. McN. (1997) A theory of mixed chains applied to safety factors in biological systems. *Journal of Theoretical Biology,* **184**, 247–52.

Currey, J. D. (Ed.) (1984) *The Mechanical Adaptations of Bones.* Princeton University Press, Princeton.

Garland, T. and Huey, R. B. (1987) Testing symmorphosis: does structure match functional requirements? *Evolution,* **41**, 1404–9.

Weibel, E. R., Taylor, C. R. and Hoppeler, H. (1991) The concept of symmorphosis: a testable hypothesis of structure–function relationship. *Proceedings of the National Academy of Sciences,* **88**, 10357–61.

# 3
# Evolution of optimal systems

## 3.1 Overview

### MALCOLM S. GORDON

All properties of living organisms are current results of evolutionary processes, some of which began with the origin of life on earth more than 3.5 billion years ago. To the extent that ideas like symmorphosis and optimization apply to organisms, the principles and processes involved, and the existing states produced, also had evolutionary origins. This chapter discusses some of the more important ways in which symmorphosis relates to and derives from evolutionary biology.

We encourage you, while reading the three essays included in this chapter, to think about several general issues:

(i) The evolutionary process has a number of basic properties:

    (a) it has neither direction nor purpose – organisms are not designed in advance;

    (b) it is entirely contingent – if it were ever possible to start any parts of evolutionary history over again from their beginnings, the chances that the results would be the same as what we see today would be very close to zero;

    (c) it is continuous and continuing – all biologists are studying moving targets changing with time;

    (d) organisms as we find them today are continuing evolutionary experiments with structural and functional properties and capacities that have resulted from both short and long term past conditions.

(ii) While natural selection is usually the principal factor that operates upon variation within organisms to produce evolutionary change, it is not the only factor. Both migrations of organisms into or out of populations and random genetic drift can also produce change that persists. Drift may be particularly significant in smaller populations. The changes produced by both migrations and drift may not be adaptive.

(iii) Even in situations where natural selection is clearly the major factor at work, that selection may work in complex ways. Many different characteristics of organisms may be selected for or against under given conditions, and selective conditions may vary greatly over both time and space. Thus the nature, direction and duration of specific selective pressures are variable and unpredictable.

(iv) Adaptive properties of organisms involve their resistances to naturally occurring ranges of ecological variables (for example, temperature, water salinity) as well as their capacities to carry on essential functions (for example, respiration, circulation). It seems unlikely that the principle of symmorphosis applies to resistances.

On the basis of these considerations one might expect that symmorphosis, to the extent that it may be found in particular circumstances, is more an accident than a principle of adaptation. It is, however, a useful and stimulating idea that can promote research work and give important insight into how organisms work.

The essays in this chapter consider the relationships between putative cases of symmorphosis in the real world and their possible evolutionary origins and backgrounds. Garland points out that the general idea of symmorphosis makes it a kind of optimality model, and that there are many good reasons for expecting that most organisms are not likely to be optimal for any single function. Symmorphotic thinking can, however, be of considerable heuristic value in efforts to understand organismal "design," and thus lead to an array of interesting and potentially informative research projects in areas such as quantitative genetics and bioengineering.

Feder points out that there are multiple ways in which experimental physiologists can test specific predictions concerning the evolutionary origins and significance of putative cases of symmorphosis. Both field and laboratory experiments testing for effects of environmental variations on animals are possible, as are actual manipulations of specific features of the animals of interest. These approaches provide opportunities for understanding evolutionary processes and dynamics.

Ricklefs summarizes some of the heuristic consequences of applying the principle of symmorphosis to the growth, development, and physical performance capacities of the skeletal muscle systems of growing birds. This effort is partly successful, and it highlights several interesting and important questions that can only be answered through additional research.

### Further reading

Blake, R. W. (Ed.) (1991) *Efficiency and Economy in Animal Physiology.* Cambridge University Press, Cambridge, UK.

Dudley, R. and Gans, C. (1991) A critique of symmorphosis and optimality models in physiology. *Physiol. Zool.,* **64**, 627–37.

Dupre, J. (Ed.) (1987) *The Latest on the Best: Essays on Evolution and Optimality.* MIT Press, Cambridge, MA.

## 3.2 Conceptual and methodological issues in testing the predictions of symmorphosis

### THEODORE GARLAND, JR.

This part of the chapter has three general purposes. First, I present some cautions about optimality models in general and about symmorphosis in particular. Second, I present the quantitative genetic perspective on correlated evolution of different aspects of the phenotype. Third, I consider how one might rigorously test symmorphosis as an evolutionary hypothesis.

### Reasons why organisms are not optimal

For various reasons, optimality models are controversial in evolutionary biology. First, organisms are not "designed," and natural selection is not engineering. Although engineers have final goals and purposes, natural selection does not. As well, engineers often design things for a single purpose, whereas organisms must do many things, not just one. In general, nature does not have the luxury of "designing" task-specific organisms. Consequently, "constraints" or "trade-offs" are pervasive in biological systems, and any sort of general or global optimality is exceedingly unlikely to occur.

Second, biological materials have limitations. Although engineers can start from scratch, natural selection cannot. Rather, selection is constrained to work with pre-existing materials (whatever a species happened to inherit from its ancestors), and these might not be the best possible materials for a particular function. For example, whales retain lungs (they have not re-evolved gills) and titanium does not occur in tortoise shells.

Third, energetic efficiency is not necessarily what selection maximizes. Often simply for convenience, optimality models are phrased in terms of energy as the common currency, and the "goal" of selection is seen as

40

maximizing net energy gain or perhaps maximizing energetic efficiency. We have little empirical evidence, however, that this is what selection actually tends to do. Instead, selection generally leads to adequacy or sufficiency, not necessarily optimality.

Fourth, environments are always changing, and selection often cannot keep pace. Selection cannot change organisms extremely rapidly for two reasons: (1) the heritability of phenotypic variation on which selection acts is usually far less than unity, especially for physiological traits; (2) if selection is too strong, then population size will be reduced such that extinction by demographic stochasticity is likely.

Fifth, even if selection could follow the pace of "typical" environmental change, it could not possibly anticipate the effects of major environmental changes, such as asteroids hitting the earth, severe droughts, "100-year floods," or even the invasion of a population by some new pathogenic organism, such as AIDS in the human population.

Sixth, genetic drift operates in all populations and can be strong enough to thwart selection. Thus, genetic drift alone should ensure that average values for populations or species are rarely if ever at the optimum dictated by selection. Indeed, genetic drift is a key element of Sewall Wright's shifting balance theory of evolution, in which it is argued that drift can often push populations in the direction of lower mean fitness.

Seventh, behavior may evolve more rapidly than physiology or morphology, leading to mismatches between what animals do and what they are best suited to do. The dipper (*Cinclus* species) is often cited as an example: it dives and forages underwater, yet is not much different from an ordinary passerine bird in terms of its morphology.

Eighth, sexual selection is largely independent of, and can act counter to, natural selection. Sexual selection, either by male–male competition for access to females or by female choice of particular males, can cause the evolution of bizarre structures and behaviors, such as the tails of peacocks, which are maladaptive with respect to natural selection.

### Optimality models can be useful tools

Even if organisms generally are not optimal, optimality models can be useful tools for understanding the evolution of physiological systems. They can indicate the best that organisms could be, given some explicit assumption of a design criterion and within specified constraints. This can facilitate quantitative tests of the degree of departure from perfect design and hence further understanding of the constraints important for the specific system being studied. Symmorphosis is an informal optimality model, so it can be a useful tool for understanding organismal "design."

## Symmorphosis is based on "common sense"

Symmorphosis is based on the notion that animals are built "reasonably," without unnecessary excess capacity – except possibly for appropriate "safety factors." Symmorphosis follows from the idea that maintenance of excess structure is equivalent to wastage of energy and/or space, and hence natural selection should tend to eliminate superfluous structures. This is a very old idea, recognized by Charles Darwin, and invoked by Bennett and Ruben (1979) in their paper discussing the evolution of endothermy: "It is reasonable to assume, however, that these coevolved transport and utilization systems will not differ greatly from each other within an individual animal in their capacity for oxygen processing." They were referring to maximal capacities for oxygen transport, delivery, and utilization, the same physiological system that motivated Taylor and Weibel's original formulation of symmorphosis.

## Constraints, trade-offs, and the quantitative genetic perspective on correlated evolution

One reason why symmorphosis might not often occur is that particular structures, and indeed entire organ systems, often must serve multiple functions. For instance, as noted by Lindstedt and Jones (1987), the skin is a (semipermeable) barrier to the external environment, is often damaged and so must be able to heal rapidly, is involved in thermoregulation, osmoregulation, and sensation, and often helps to camouflage an organism. Can such an organ possibly be "optimal" in any meaningful sense? Similar arguments would apply to the respiratory system, which, in addition to its primary function of oxygen transport and delivery, must also get rid of $CO_2$ and, in many organisms, be involved in thermoregulation and sometimes sound production.

Interactions of the elements within complex systems are often discussed under the general rubric of "constraints" or "trade-offs." One often-discussed example involves possible trade-offs between locomotor speed and endurance: for a given mass of muscle, increasing speed of contraction may come at the price of decreased fatigue resistance.

The framework of quantitative genetics provides one way to think about these kinds of complexities. This branch of genetics was developed for quantitative phenotypic characters that are polygenic (affected by many genes), such as body size or metabolic rate. It provides equations to predict the response to selection, be it natural, sexual or artificial.

For a single character, the equation to predict the evolutionary change of a population's mean phenotype from one generation to the next is very simple: $R = h^2 S$. Thus, if we know what character selection ($S$) is favoring at the phenotypic level, and the narrow-sense heritability of that trait ($h^2$, which ranges from 0 to 1), then we can directly predict both the direction and rate of evolutionary change in the phenotype ($R$).

Organisms comprise far more than just one character. The multivariate version of the above equation is far more complicated: $\Delta \bar{Z} = G P^{-1} S$, where $\bar{Z}$ is the vector of phenotypic means for a series of traits, $G$ is the additive genetic variance–covariance matrix (reflecting narrow-sense heritabilities and genetic correlations), $P$ is the phenotypic variance–covariance matrix, and $S$ is a vector indicating the apparent selection acting on each trait.

Consequently, for multiple traits, the direction and rate of response to selection cannot be accurately predicted in the absence of fairly complete information regarding the selection affecting, and the genetics of, all traits that are correlated with the trait of primary interest. Moreover, many empirical examples indicate that responses to artificial selection are often unpredictable, especially with regard to traits other than the one(s) of primary interest. Accordingly, prediction of the pathway of phenotypic evolution is not easy, and real organisms often may fail to match our "common-sense" expectations.

## A graphical model of symmorphosis

Taylor and Weibel argued specifically that when a physiological system is working at its maximum capacity, all steps within the system should be at their maximum, such that none is in excess and so none is, by itself, limiting to overall flux. A simple diagrammatic model of this idea is presented in Figure 3.1. Symmorphosis claims that selection should eliminate "excessive construction," resulting eventually in organisms that correspond to the case on the right of Figure 3.1, in which the capacity of each step is perfectly matched. This state can also be called highly "integrated." An organism fitting the pattern depicted on the right of

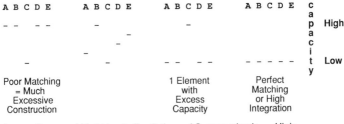

44         *T. Garland, Jr.*

Figure 3.1. A diagrammatic model of symmorphosis, illustrating four possible cases of the matching of capacities of elements (A–E) in a physiological chain. In all four, the maximal capacity of the physiological system (for example, oxygen transport, delivery, and utilization) is the same. In the rightmost case, maximal flux through the system is limited simultaneously by all elements (or, equivalently, it is limited by none). In the leftmost case, flux through the system is limited only by element C, which has the lowest maximal capacity; all of the other elements exhibit "excessive construction". Organisms resembling the case on the left would have "wasteful" excess capacities and hence, symmorphosis predicts, they should rarely exist in nature. Similar to the third case from the left, the mammalian respiratory system often seems to exhibit excess lung capacity, with capacities of other elements fairly closely matched.

Figure 3.1 would be completely symmorphotic, or optimal in some sense. (Note that symmorphosis does not specify how high the maximal capacity of a physiological system should be; only that whatever that maximal capacity (as dictated by selection), all components should be just sufficient in their capacities.) An organism fitting any of the left three patterns depicted in Figure 3.1 would be suboptimal in the sense specified by symmorphosis.

Given such a model of symmorphosis, it should be possible to devise a system to assign a quantitative score for each species or individual animal that one might measure, indicating the extent of wastefulness summed across all of the excessive elements. Supporters of symmorphosis as a generally useful description of animals in nature would expect to find the frequency distribution depicted in Figure 3.2 (top left), with most animals being perfectly symmorphotic (all elements matched in their maximal capacity) or nearly so. The opposite possibility is that most organisms are far from optimal (Figure 3.2, bottom left). If, on the other hand, most organisms are "adequate" or "sufficient," then the frequency distribution of organisms in nature might look something like what is shown in the top right of Figure 3.2. Another empirical possibility is that nature has produced a broad array of organismal diver-

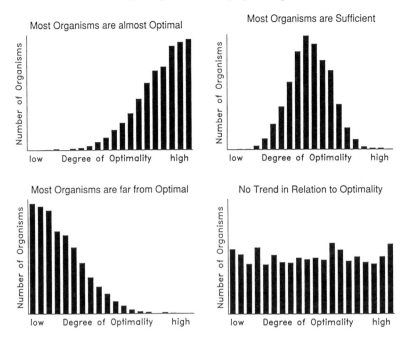

Figure 3.2. Four possible alternatives for the frequency distribution of organisms in nature. The horizontal axis represents the degree of approach to the condition of symmorphosis (defined as close matching of the capacities of components within a physiological system, as depicted on the right of Figure 3.1). Symmorphosis predicts that most organisms are at least nearly symmorphotic (top left), whereas most evolutionary biologists would probably predict that either the top right or perhaps the bottom right is closer to the truth. We currently lack sufficient data to compile such a figure for real organisms, but the significance of symmorphosis as an evolutionary principle will depend on the relative frequency with which its predictions are matched by real organisms.

sity, with some organisms being optimal, others rather poorly "designed," and all possibilities in between (Figure 3.2, bottom right).

Whether most organisms are only crudely integrated, sufficient (adequate), or nearly optimal (perfect) is an empirical question. Tests of symmorphosis by Weibel, Taylor, and their colleagues, with a variety of mammalian species, have produced results that do not indicate perfect matching (rightmost case in Figure 3.1). Specifically, the lung often seems to have excess capacity. As of yet, however, we lack sufficient data to determine the generality of such results for any physiological system in any group of organisms.

## A cautionary note about "safety factors"

Most biologists agree that organisms often possess capacities somewhat in excess of what they typically use. Animals, for example, walk most of the time, but they have bodies that can (usually) take the forces experienced during sprint running. From this observation, biologists infer that events which occur only rarely during the lifetimes of individual organisms can nonetheless constitute important selective factors.

Because the concept of safety factors is essentially derived from engineering, whereas organisms are not "engineered," we should be careful about this metaphor or analogy. We should also be very careful in calling *any* excess capacity a safety factor – with the implication that the safety factor is probably optimal in some sense – or else we may fall into the trap of "foregone confirmations", as suggested by Lindstedt and Jones (1987). For example, Karas *et al.* (1987) suggest that "The excess diffusing capacity can be thought of as a 'safety factor' allowing animals to tolerate a decrease in $PAO_2$ or $DLO_2$ that may occur normally at altitude or during disease, but it is not clear why this factor should be smaller in the more athletic species." This kind of argument comes perilously close to representing a "just so story" of the kind rightly criticized by Gould and Lewontin (1979). For instance, most organisms (unlike some expeditionary physiologists!) live their lives within a limited altitudinal range, such that symmorphosis would seem to actually predict an elimination of unnecessary "safety factors."

We must be very cautious in accepting claims that every instance of apparent excess capacity represents a "safety factor" (with the implication that it is maintained by natural selection, even though we may admit ignorance as to exactly why) or else we risk tautology and failure to better understand nature. In fact, in no case has a putative "safety factor" in vertebrate physiology or morphology been empirically demonstrated to exist for the implied purpose.

## Further reading

Arnold, S. J. (1992) Constraints on phenotypic evolution. *American Naturalist*, **140** (Supplement), S85–107.

Bennett, A. F. and Ruben, J. A. (1979) Endothermy and activity in vertebrates. *Science*, **206**, 649–54.

Boake, C. R. B. (Ed.) (1994) *Quantitative Genetic Studies of Behavioral Evolution*. University of Chicago Press, Chicago.

Dudley, R. and Gans, C. (1991) A critique of symmorphosis and optimality models in physiology. *Physiological Zoology*, **64**, 627–37.

Dupre, J. (Ed.) (1987) *The Latest on the Best: Essays on Evolution and Optimality.* MIT Press, Cambridge, MA.

Feder, M. E., Bennett, A. F., Burggren, W. W. and Huey, R. B. (Eds.) (1987) *New Directions in Ecological Physiology.* Cambridge University Press, New York.

Garland, T., Jr. and Adolph, S. C. (1994) Why not to do two-species comparative studies: limitations on inferring adaptation. *Physiological Zoology,* **67**, 797–828.

Garland, T., Jr. and Carter, P. A. (1994) Evolutionary physiology. *Annual Review of Physiology,* **56**, 579–621.

Gould, S. J. and Lewontin, R. C. (1979) The spandrels of San Marco and the Panglossian paradigm: a critique of the adaptationist programme. *Proceedings of the Royal Society of London B,* **205**, 581–98.

Jacob, F. (1977) Evolution and tinkering. *Science,* **196**, 1161–6.

Karas, R. H., Taylor, C. R., Jones, J. H., Lindstedt, S. L., Reeves, R. B. and Weibel, E. R. (1987) Adaptive variation in the mammalian respiratory system in relation to energetic demand. VII. Flow of oxygen across the pulmonary gas exchanger. *Respiratory Physiology,* **69**, 101–15.

Lindstedt, S. L. and Jones, J. H. (1987) Symmorphosis: the concept of optimal design. In *New Directions in Ecological Physiology*, pp. 289–309. Eds. Feder, M. E., Bennett, A. F., Burggren, W. W. and Huey, R. B. Cambridge University Press, New York.

Parker, G. A. and Maynard Smith, J. (1990) Optimality theory in evolutionary biology. *Nature,* **348**, 27–33.

# 3.3 Testing the evolutionary origin and maintenance of symmorphosis

MARTIN E. FEDER

Many studies of symmorphosis assume that the equilibration of functional demand and functional capacity results from natural selection, in which organisms that have an appropriate functional capacity leave more offspring than organisms with either too much or too little functional capacity. However, processes other than natural selection (for example, sexual selection, genetic coupling, genetic drift, and constraint) can also potentially yield the same outcome. Furthermore, regardless of the origin of structure–function matching, its present maintenance could be due to mechanisms other than natural selection. How can we distinguish among these alternative explanations and evaluate the relative contributions of each potential evolutionary mechanism to the origin and maintenance of the present phenotype? Evolutionary biologists have developed powerful theoretical and experimental tools that physiologists can exploit to answer these questions. My purpose is to describe some of these tools and how they might be used to greatest effect.

## How to examine natural selection and its alternatives experimentally

Both natural and artificial or experimental selection involve a screen in which some individuals in the study population are excluded from reproducing or are caused to reproduce at a lower rate than other individuals in the population (Figure 3.3). The screen can be by outright death, by sublethal effects that inhibit reproduction, or by culling individuals before reproduction. For example, an experimentalist interested in aerobic metabolic capacity might measure the maximum rate of oxygen consumption of each individual in an initial population, and breed only those individuals that score in the upper 10 percent. The entire procedure is then repeated upon the offspring of each successive generation. Measuring

48

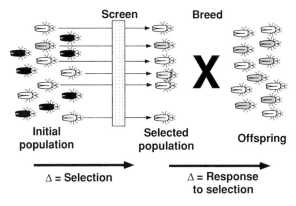

Figure 3.3.   Components of a selection experiment. A study population is sub-jected to an experimental screen, by which not all individuals survive or are able to reproduce. The difference between the initial population and those remaining after the screen indicates selection. The difference between the initial population and the offspring of those remaining after the screen indicates the response to selection in the next generation. In this cartoon, members of the initial population differ in coloration, and the screen selects for light-colored individuals. Such differences between those remaining after the screen and their offspring may be due to a variety of causes, including low heritability of coloration (see Figure 3.4).

individuals of each generation will yield the rate of change of each char-acter under study. In selecting for a high aerobic metabolic capacity, for example, one might be interested in the oxygen fluxes at each step of the pathway for oxygen from the atmosphere to the mitochondria. Symmorphosis predicts that these fluxes should be similar, with no excess capacity at any step. Determination of each flux for each generation, while an overall increase in aerobic metabolic capacity evolves, will reveal whether all component fluxes evolve in tandem or whether certain fluxes outpace the others. Of course, a possible outcome is that little or no evolution occurs, with the mean maximum oxygen consumption the same for each successive generation. Such experiments need not involve an exotic or extreme stress to address the evolutionary origin and main-tenance of symmorphosis. Symmorphosis predicts, for example, that indi-viduals with too much or too little of a given trait should leave fewer offspring than individuals with exactly the required functional capacity (that is, stabilizing selection). One can compare the distribution of a functional capacity (for example, pulmonary diffusing capacity, total perfusive conductance) in a population and in the offspring of that popu-lation to determine how frequently extreme individuals occur and whether extreme individuals actually leave fewer offspring.

This highly simplified scenario omits numerous details and pitfalls. Some of the relevant considerations are as follows:

- Experimental studies of the response to selection may require measurements of at least two generations. Studies of populations in the wild may require that both parents and offspring be accessible to measurement and not be affected by the measurements. Laboratory studies may require that subject species breed in the laboratory and offspring grow to an age at which measurements can be made. These requirements constrain the choice of model systems.
- Practitioners of selection studies and artificial selection have developed an extensive body of theory and practice with which to carry out and interpret such studies. This knowledge includes robust statistics for characterizing parent and offspring populations and any differences between them, breeding designs that yield the largest and least ambiguous outcomes for the smallest number of crosses, and theoretical models that aid distinction among natural selection and its alternatives. Some key components include the distinction between selection proper and the response to selection (Figure 3.3). If a trait (for example, metabolic rate, pulmonary diffusing capacity) is measured on both an initial population and its collective offspring, the measurements of offspring can be regressed against those of parents (Figure 3.4) to yield a slope (representing heritability) and a correlation coefficient. These measures are important because, even if selection is significant, the response to selection in the next generation may be negligible if the heritability is low.
- Phenotypic plasticity, an organism's ability to alter its phenotype within its lifetime, can confound studies of selection unless properly controlled. Examples of phenotypic plasticity include training, acclimation, instances in which the rearing of individuals in differing environments results in differing but fixed adult phenotypes, and so on. Many of the traits studied in reference to symmorphosis are phenotypically plastic (see, for example, the work of Diamond in Chapter 2.3).

### Useful paradigms for potential exploitation

Evolutionary biologists, plant and animal breeders, molecular biologists, and Nature itself have developed many interesting variations on these

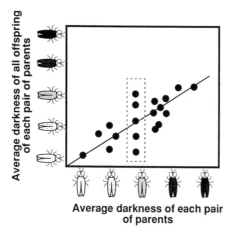

Figure 3.4.   A possible relationship between coloration of parents and their off-spring. The slope of the least-squares regression line is equivalent to the herit-ability. Note that in this hypothetical example, parents of a given darkness can produce offspring varying greatly in darkness (see points within the broken rec-tangle). Given such effects, even if selection is strong, the rate of evolution can be slow.

themes. Those interested in understanding the evolution of symmorpho-sis can exploit this variation.

### *Response to an experimental environment*

Placing a population in a defined environment (hot or cold, constant physical activity versus quiescence, high elevation versus sea level, and so forth) has the advantage of simulating the way in which natural selec-tion is thought to act. A disadvantage, however, is that the specific evolu-tionary response to selection is unpredictable. The raw variation upon which selection works is random; each selective screen will operate upon only the particular variation that happens to be present, and thus multi-ple evolutionary outcomes will ensue. For example, many different organic compounds can and do serve as proton acceptors in adaptation to hypoxia or anoxia. Of course, symmorphosis predicts that whatever response evolves, none of its components should exceed the others. The following possibilities represent a spectrum from greater to lesser eco-logical realism and relevance, and from lesser to greater experimental control.

## Natural selection in the wild

Free-ranging organisms can be examined in the wild, where the experimental screen comprises the stresses that these organisms naturally encounter. Two salient examples are the work of Stevan Arnold (University of Chicago), Albert Bennett (University of California, Irvine) and colleagues on garter snakes (*Thamnophis*), and the work of Ward Watt (Stanford University) and colleagues on butterflies (*Colias*). In the snake work, juveniles are marked, measured and their physiological capacities for burst and sustained activity determined; snakes are then released and recaptured at later times. From these repeated measurements, the investigators can deduce whether interindividual variation is consistent through time, the relationship between particular traits (excessively long tails or high stamina) and survival, and, in theory, the relationship between these traits and reproductive success. Such information is combined with laboratory studies of heritability of the traits and theoretical models of evolution to predict the response to selection in the next generation. This system could equally well be used to examine the impact of excess capacity in studies of symmorphosis.

The butterfly work examines natural populations with multiple alleles for phosphoglucose isomerase, an enzyme of glycolysis. Some alleles enhance metabolic flux at typical temperatures, while others are advantageous primarily at warm temperatures; metabolic flux, in turn, is linked to flight performance and courtship ability. One issue is whether the less frequent but significant advantage of the warm alleles is sufficient for natural or sexual selection to maintain it in natural populations. Sampling of allele frequencies in the natural populations over time can elucidate this and other issues.

## Pre-existing experiments

Natural selection in the wild can be slow and ambiguous. Thus, investigators can benefit when selection has been operating for many generations before a study begins and where the selective agent is unambiguous. Such instances include accidental or intentional introductions into extreme environments, and artificial selection by plant and animal breeders. Ideally, studies should measure traits at regular intervals (for example, every generation or every other generation) because the kinetics of evolution or stasis can elucidate underlying evolutionary mechanisms. Fortunately, both introductions into extreme environments and breeding programs often are well documented. One classical example is the introduction of humans and domestic animals to high-altitude environments. Investigations of capacity–demand matching as a function of the number of generations spent at high elevation could contribute to understanding the evolution of symmorphosis.

### De novo experiments

Finally, the investigator can initiate selection in either the laboratory or the field, but directly control the selective screen by experimentally imposing known stresses upon the subject populations and/or by choosing individuals for breeding on the basis of some predetermined criterion. The obvious problem of such experimentation, that selection requires many generations, can be mitigated by choosing subjects with short generation times.

Most potential and actual experiments impose an increased load or stress upon the experimental population. A potentially insightful variant of this procedure would be to reverse it: to take a population that has undergone selection in an extreme environment and transfer it to a benign environment. At the time of transfer, the relevant functional capacities would presumably be in gross excess of the relatively modest demands of the benign environment. According to the expectations of symmorphosis, natural selection ought then to result in the abolition of excess capacity.

### *Experimental manipulation of a trait*

As explained above, an expectation of a selection experiment is that *some* traits should evolve in response, but *which* traits evolve is largely a matter of chance. Selection for increased locomotor stamina could result in neural changes that facilitate more efficient locomotion, enhanced ability to supply ATP to the locomotor musculature, both, or neither. Thus an alternative approach is to manipulate a chosen trait directly: increase body size, decrease bone strength, change hematocrit, and so on. When the manipulated traits are part of multicomponent systems, symmorphosis predicts that natural selection should operate to bring their associated capacities into register with the capacities of other components of the system. When the manipulation yields capacities greatly in excess of that required in a given environment, symmorphosis predicts that natural selection should operate to eliminate the excess capacity.

A straightforward way to manipulate single traits is through the classical techniques of physiology and experimental morphology: delete, diminish, augment, or add traits through surgery, pharmacological intervention, and so on. These need not be elaborate; weights could be added to experimental animals or supernumerary limbs removed for generation after generation. The "allometric engineering" of Barry Sinervo (University of Indiana) and Raymond Huey (University of Washington), in which they manipulated the size of hatchling lizards by surgically excising yolk of some littermates but not others, is an excellent example of this approach. Importantly, the manipulation needs to be performed on successive generations for selection to continue, inasmuch as the manipulation (as opposed to the responses to it) will not be inherited.

Alternatively, genetic manipulations can be inherited by subsequent generations without re-application. Students of symmorphosis have often not exploited this approach, perhaps because many genetic variants were not especially relevant to symmorphosis. An exception was the work of Ewald Weibel on waltzing mice, which bear a mutation that causes sustained locomotion. Advances in molecular genetical techniques, however, are yielding numerous mutants that are superb raw material for studies of symmorphosis and selection. Most of these mutations have been studied for their own sake; few have been studied with respect to if and how selection alters background genes to cope with the changes in demand or capacity conferred by the mutant genes.

## Conclusion

As comparative physiologists intensify their examination of evolutionary issues and principles, they increasingly need to understand mechanisms of evolutionary change. No self-respecting physiologist would measure a blood pressure, a contractile force of a muscle fiber, an ion concentration within a fluid compartment, or a membrane potential for a single point in time and stop there, for any single point measurement is the product of multiple interacting processes whose dynamics need to be understood to comprehend the significance of the single point measurement. Similarly, the physiological and morphological phenotypes that physiologists study are but single points in evolutionary time, and the testing of evolutionary hypotheses such as symmorphosis that seek to explain the genesis of these phenotypes requires that we understand the dynamics of the interacting evolutionary forces that produce these phenotypes. On the one hand, the study of evolutionary dynamics may require sustained interaction with an unfamiliar group – evolutionary biologists conversant with the experimental study of natural selection and its alternatives. On the other hand, the elucidation of evolutionary dynamics often requires that symmorphologists do not discard their existing research paradigms and programs, but expand them to examine different ages or generations of experimental subjects. Much is to be gained by exploiting this new direction.

## Acknowledgments

Supported by NSF Grant IBN94-08216 and the Louis Block Foundation.

## Further reading

Bennett, A. F. and Huey, R. B. (1990) Studying the evolution of physiological performance. In *Oxford Surveys in Evolutionary Biology*, vol. 7, pp. 250–84. Eds. D. J. Futuyma and J. Antonovics. Oxford University Press, New York.

Feder, M. E. and Watt, W. B. (1993) Functional biology of adaptation. In *Genes in Ecology*, pp. 365–91. Eds. R. J. Berry, T. J. Crawford and G. M. Hewitt. Blackwell Scientific Publications, Oxford.

Feder, M. E., Bennett, A. F., Burggren, W. W. and Huey, R. B. (Eds.) (1987) *New Directions in Ecological Physiology*. Cambridge University Press, Cambridge.

Rose, M. R. and Lauder, G. V. (Eds.) (1996) *Adaptation*. Academic Press, New York and London.

## 3.4 The concept of symmorphosis applied to growing birds

ROBERT E. RICKLEFS

The basic idea of symmorphosis is one of economy of design, specifically that interrelated components of a system are mutually adjusted so no single component has a capacity in excess of others. In this contribution, I ask whether the concept of symmorphosis can help us to understand the design of the growing organism. The answer to this question is unlikely to be straightforward because (1) the growth of different structures produces parallel demands on resources available to the organism; (2) performance is difficult to define; and (3) the costs and benefits of variation in performance are difficult to evaluate in terms of increments of evolutionary fitness.

Two measures of performance are important to the survival and fitness of young organisms. One is growth, the other is self-maintenance. To some extent, these functions are mutually incompatible. The organism comprises many different structures that begin to function in a mature manner at different times in the course of its development. To establish whether growth and maturation of different structures are symmorphic, one must find a way to match their capacities to the performances that are required of them. The degree of matching is only properly assessed by the fitness consequences of not being matched, which unfortunately cannot be measured in most systems. Nonetheless, we may gain some insight into constraints on the design of the organism by comparing the growth and development of different organs.

The analyses presented here are based on studies of growth in birds, particularly the growth of skeletal muscles, which vary considerably in the onset of their function. In precocial species leg muscles function actively from hatching, but in altricial species, their function increases at a later age. Pectoral flight muscles begin to function early in some precocial species, whereas in other precocial species and in altricial species, they develop function only after they have achieved most of their adult mass.

The problem that birds must solve is how to grow different structures of the body in such a way that they deliver needed function during the course of development and yet attain adult size within a reasonable period. In principle, the concept of matching functional and design parameters is directly applicable to the problem of postnatal growth. However, these concepts differ depending on whether one views growth as limited by the supply of materials and metabolizable energy to growing tissues or by intrinsic capacities of tissues themselves for growth. In a supply-limited system, different tissues compete for resources and the growth strategy should optimize the allocation of resources among tissues with respect to the need for their function at different times during development. In a system limited by intrinsic capacities of tissues, growth may be incompatible with functions characteristic of mature tissues (mature function). In this case, an inverse relationship between growth rate and mature function of tissues may be a basic functional parameter of growing systems.

Each structure of the organism presumably is constrained by its own functional parameters. How, then, should each structure grow and mature so that the organism as a whole develops in some evolutionarily "optimized" manner. Greater maturity at a given size presumably enhances survival rate, but prolongs the growth period.

Different structures begin to function at different times during the growth of the individual. It makes sense that those structures that are called on early in development should attain large size quickly and mature at an early age, after which their rate of growth should decrease. Structures that are called on later in development presumably should delay maturation, and initially could be small relative to adult size because of their greater growth capacity. In general, this expectation is borne out when the structures of avian neonates are compared. In the neonates of precocial birds, the brain, gut, and leg muscles, which are relatively matured, as they must be for the chick to feed itself soon after hatching, are larger proportions of their adult sizes than are the flight muscles, which will not be called on to do their job for several weeks.

How can we use the principle of symmorphosis to study the design of the growing chick? One possibility is to compare the growth and maturation of several structures among species and ascertain the degree to which the growth patterns of two or more structures vary in parallel. Such a comparison would be particularly revealing if the ontogeny of function of one of the structures varied among species and that of a second did not. Most species of bird do not begin to fly until nearly fully grown, whereas the ontogeny of walking varies tremendously from the cursorial neonates of precocial species to the helpless chicks of altricials.

If the growth patterns of different structures were adjusted with respect to one another so that their growth periods were maintained in similar relationship, one would expect to find mutual adjustments in the relative sizes or levels of functional maturity of the structures. This proposition is evaluated here by examining postnatal growth increments and an index to functional maturity at hatching of the flight muscles and the muscles of the legs of a variety of species of bird. Three measures of each of the muscle masses will be used: (1) the postnatal growth increment, which is the logarithm of the ratio of the masses of the muscles in the adult and neonate; (2) the dry matter contents of the muscles of the neonate, which provide indices to the capacity of the muscle for mature function (Ricklefs and Webb 1985; Ricklefs, Shea and Choi 1994); and (3) the rate constants of Gompertz growth models fitted to each of the muscle masses, which measure the rate at which muscles approach their mature sizes. In addition, the ratio of the growth increment (dimensionless) to the growth rate constant (1/days) provides an index to the length of the postnatal growth period (days), which is one measure of growth performance.

### Performance of growing muscles

The relative sizes of the leg and pectoral muscles in neonates and adults exhibit considerable variability among species related to locomotion (Figure 3.5). The pectoral muscles of neonates are, for the most part, smaller than the combined muscles of the leg. The situation is reversed among the adults, in which the masses of the pectoral muscles generally exceed those of the leg muscles. The data suggest that the muscle masses of adults exceed those of neonates by one to two orders of magnitude for the leg muscles and two to three orders of magnitude for the pectoral muscles. The growth increments of the two muscle masses, compared in Figure 3.6, bear this out.

### Comparative analyses

The proportion of nonaqueous material, or the dry fraction (DF), is correlated with functional maturity of skeletal muscle. Values of the

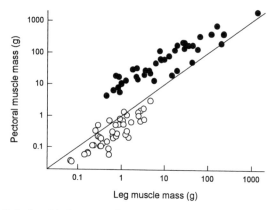

Figure 3.5. Relationship between the masses of the pectoral and leg muscles in adult (solid symbols) and neonate (open symbols) birds.

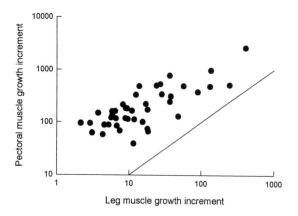

Figure 3.6. Relationship between the postnatal growth increments (adult/neonate mass) of the pectoral and leg muscles. Symbols as in Figure 3.5.

DF of the leg and pectoral muscles are consistent with variation in the functional maturity of neonates. For example, the dry fractions of the altricial species in the sample (0.11–0.16) are considerably lower than those of the precocial ducks (0.23–0.24), wading birds (0.19–0.25), and galliforms (0.16–0.19), among which differences in leg muscle dry fraction are consistent with differences in the thermoregulatory abilities of the chicks (Visser and Ricklefs 1993). The semiprecocial larids (0.18–0.23) and alcids (0.18–0.20) are similar to precocials; petrels (0.15–0.24), especially the smaller species, have high leg muscle dry fractions to support their formidable thermoregulatory abilities.

The relationship between the growth periods (increment/growth rate constant) of the pectoral and leg muscles is shown in Figure 3.7. The growth periods of the pectoral muscles generally exceed, often considerably, those of the leg muscles, and the correlation between the two is weak. The growth periods of the leg muscles in some groups of birds (ducks, larids, shorebirds, plovers, and some petrels) are less than 50 days, and mostly less than 30 days, yet the growth periods of the pectoral muscles of these species are between 50 and 150 days. This would appear to be a gross mismatching of the performances of different organs in the same individual.

## Discussion

With respect to the growth of the leg and pectoral muscle masses of birds, the idea of symmorphosis might be revealed by mutual adjustments of the designs of structures under strong constraint, in which case variation in the performance of two or more structures could reduce the fitness of the individual. Two premises of this analysis were that an inverse relationship between growth rate and mature function is a basic functional parameter, or design constraint, of muscle, and that the growth period of a structure influences fitness, particularly when the growth periods of different structures vary.

One design feature consistent with the concept of symmorphosis is the small relative neonatal size of the pectoral muscles and their high capacity

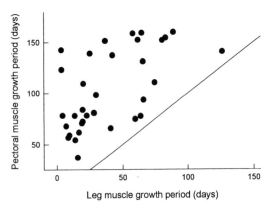

Figure 3.7.   Relationship between the growth periods (postnatal growth increment/Gompertz growth rate constant) of the pectoral and leg muscles of several birds. Symbols as in Figure 3.5.

for postnatal growth. Pectoral muscles of neonates are relatively large in species that fly at an early age (galliform birds) or that use their pectoral muscles for thermogenesis early in postnatal development because their small legs have limited capacity to generate heat (ducks and petrels). Both the pectoral muscles and the leg muscles of neonates are relatively small in some altricial birds in which the onset of mature function of both muscle masses is delayed compared to precocial species. The pectoral muscles have a higher growth potential throughout most of the postnatal growth period, which is consistent with their relatively higher exponential growth rates and lower dry fractions at hatching, and initially a slow increase in dry fraction with respect to increasing size. More rapid growth partly balances the greater growth increments of the pectoral muscles. However, pectoral muscles require much longer to complete their growth than do the muscles of the legs. This has the effect of delaying the attainment of adult body proportions and perhaps the onset of flight.

How do these analyses bear on the idea of symmorphosis? First, growth is a complex phenomenon. The "performance" of the growing organism must be measured in terms of both its rate of growth and its function at each point during the growth period. In the case of pectoral and leg muscles, dry fraction is probably an inadequate index of function. Furthermore, in order to evaluate the hypothesis of a growth rate–mature function trade-off, one may have to use more detailed comparisons than the correlation between growth rate constant of fitted equations and the growth increment and initial dry matter fraction. The examination of growth in the context of symmorphosis is made all the more difficult because growth rate need not be pushed to the limit set by the intrinsic quality of tissue (that is, its level of function) but may be limited by other factors not included in the present analysis.

This leads to the second point, which is that the organism is made up of many structures which have various degrees of interrelationship. It might be hoping for too much to expect a close correspondence between the performances of any two structures. It would seem more prudent to search for patterns of correspondence between structures, either in growth performance or in functional performance, that revealed the overall design features and functional parameters of the architecture of the growing organism.

Finally, symmorphosis probably should not be considered as a testable hypothesis. Rather, it is a guiding principle for investigations into the architecture of organisms. One wishes to match capacity and performance of the structures of the organism. Presumably, unused capacities and poorly matched performances reduce the fitness of the organism, but they also can reveal constraints on design and help us to understand how the response to selection is limited. The growing organism represents a particularly challenging system for the study of organism architecture because aspects of performance relevant to selection include both function and growth.

## Further reading

Carrier, D. and Auriemma, J. (1992) A developmental constraint on fledging time of birds. *Biol. J. Linn. Soc.*, **47**, 61–77.

Ricklefs, R. E. (1979) Adaptation, constraint, and compromise in avian postnatal development. *Biological Reviews*, **54**, 269–90.

Ricklefs, R. E. and Webb, T. (1985) Water content, thermogenesis, and growth rate of skeletal muscle in the European Starling. *Auk*, **102**, 369–76.

Ricklefs, R. E., Shea, R. E. and Choi, I. (1994) Inverse relationship between functional maturity and exponential growth rate of avian skeletal muscle: a constraint on evolutionary response. *Evolution*, **48**, 1080–8.

Sibly, R. M. and Calow, P. (1987) Growth and resource allocation. In *Evolutionary Physiological Ecology*, pp. 37–52. Ed. P. Calow. Cambridge University Press, Cambridge.

Starck, J. M. (1993) The evolution of avian ontogenies. *Curr. Ornithol.*, **10**, 275–366.

Visser, G. H. and Ricklefs, R. E. (1993) Temperature regulation in neonates of shorebirds. *Auk*, **110**, 445–57.

# 4

# Bone design and biomechanics

## 4.1 Overview

SIMON MADDRELL

Bone is such a fundamental component of vertebrates that it is easy to overlook its importance. Yet the skeleton fulfills many functions and it affects and is affected by many other systems. Its coadjustments with these systems is a potent factor affecting the finished state of all these.

Of course, bones provide support for the body and allow muscles to move all or parts of the body in a great variety of ways. Bone tissue varies a great deal – in both composition and disposition – to allow it to provide different mechanical properties in different positions in the body and in different parts of itself. It has to respond to many types of loading such as compression, tension, bending, and torsion. The concept of symmorphosis suggests that structures be matched to functional requirements to optimize design at all levels of organization. This idea forces one to look very widely to secure explanations for adaptations that look maladroit or even counter-intuitive when examined from a narrower point of view. We shall see how this casts light on skeletal safety factors in relation to size, on how bone design relates to strain distribution and the need to be remodeled, and on how muscle–tendon systems attached to bone need to be constructed to provide both elastic energy saving and control of movement.

Finally, we shall see how bone continuously adjusts to change in the loads placed on it. These adjustments include both hypertrophy in response to increased load and changes in the nature of the bone. Bone remodeling that results in changes in property without change in extent are particularly appropriate where space is limiting or where to add extra bone would incur large penalties from the increased mass. Changes in nonloaded bones elsewhere in the body in animals stressed by loading raises intriguing questions as to the mechanism of adaptive response of bone to different forces.

# 4.2 Optimality in the design of bony elements
## PIETER DULLEMEIJER

The optimality problem is connected to the question of the relationship between structure and activity. The problems and difficulties met in proving optimality and the consequences are demonstrated for the biological optimality concept in the relationship between bone structure and bone activity. This relationship can be approached from two directions; that is, starting from a specific bone structure and deriving the optimal function, or starting from the function and deriving the optimal bone shape. I follow the latter approach because it shows more clearly the methodological problems involved.

Bone serves many functions, but for simplicity I shall concentrate on the mechanical function and consider the other functions as boundary conditions for the mechanical structure or the other functions as subordinate to mechanics.

### Deduction of bone shape and structure

In constructing models for bone shape and structure that meet these demands we must deal with the variables mentioned in Table 4.1. Action forces arise from muscle activity and weight-bearing. What we need for the estimation of optimality is the maximum value of the forces under normal conditions, which is not necessarily equal to the sum of the forces of all the muscles, because the organism is equipped with a feedback regulating mechanism which can diminish or balance the force. Similarly the weight of the whole animal can be carried by one, two or four legs in a quadruped, with or without any weight carried by the animal.

Reaction forces are more difficult to estimate than action forces. They are affected by the resistance of the medium in which the construction

Table 4.1. *Functional demands on bone structure*

1. Movability
2. Room for blood vessels, nerves, etc.
3. Attachment of muscles, connective tissues, etc.
4. Resistance to strain
   A. Diagram of forces
      1. Action forces
         a. muscle structure
         b. muscle attachment
         c. muscle activity (force)
      2. Reaction forces
         a. supports (structure of joints and capsules)
         b. type of construction (stabile, labile, statically determinate, etc.)
         c. position of the element
   B. Stresses
      1. Type of stress
      2. Type of material
      3. Type of construction
   C. Amount of strain allowed
   D. Optimal design

moves, the weight of the members in the construction, and the load that these members must bear.

The loading pattern can be calculated only when the direction and the magnitude of all action forces and reaction forces are known exactly. These conditions are met only when the construction is statically determinate; that is, when the number and positions of the reaction forces counterbalance precisely, and in only one way, the number and position of the known action forces. If there are many possible positions of a reaction force in cases in which only one is needed, the actual position at any moment will vary; consequently, the magnitude of the reaction force can vary. When there is more than one potential reaction force at such a spot, the construction becomes statically indeterminate.

Once we have determined the force pattern, including all the uncertainties and variabilities, we can ask which bone shape, structure and composition would be the optimal solution to resist the strain that would arise from this force pattern. A first assumption here could be that bone tissue has only one mechanical property expressed in the modules. This assumption is, however, very unlikely because many types of strains such as compression, tension, bending, and torsion will occur, and for each a particular substance will have a different modulus. An optimal solution would therefore imply that at each spot the material would be adapted to the kind of strain and stress likely to occur. This happens indeed to a certain degree. Bone tissue is a composite material. Animals have evolved a wide variety of structural materials that resist forces to protect and frequently to hold the soft tissues in place, and to shape or support them. These materials generally consist of several components, including one for resisting tensile forces, one for resisting compressive forces, and a third one which cements the other two together. These components are typically arranged and interwoven in a very complex manner.

### Bone fine structure and mechanical strength

For instance, to obtain the maximum resistance to compression, a bony element should consist exclusively of closely packed crystals with a very high Young's modulus, somewhere in the order of 100 GPa. But even so, a compromise must be struck between the mass to be carried along and resistance to compression (Alexander 1985). The disadvantage of such a construction lies in the low resistance of these materials against tensile forces. Long polymers of protein or carbohydrate are much better suited for this latter purpose. Protein polymers probably constitute the best material available in nature, although its strength is only about 0.9 GPa (Alexander 1985). From the viewpoint of an engineer, this is surprising, and the more so as in plant materials we find much stronger fibers; in wood they may be 100 times stronger.

It is very likely that collagen, of all the proteins that could be made by the organism, had the best potential, as it has suitable mechanical properties, combined with relatively easy assembly from amino acids and thus also an easy turnover, breaking down into subunits used in growth and differentiation.

It has a strong capacity for energy storage and, last but not least, it has a maximum damping capacity for vibrations exactly in the range of frequencies of nerve signals which stimulate muscle contraction (Ramaekers 1978). Collagen forms a very good mesh for the deposition of apatite crystals. Of course, this could not have been a factor in the origin of collagen, but it has made possible the development of bony tissues in vertebrates. This key innovation was followed by an enormous radiation of structural types of bone.

The polymer form of collagen is a specific clockwise helical structure, with subunits that are themselves helices. This strengthens the fibers considerably. In bone, the collagen fibers resist compression because they are packed between crystals with twisted string connections to the crystals. Also, all fibers perpendicular to the compressive stress are under tensile strain and so help indirectly to withstand the compression.

To resist compression or tension, the cross-sectional area of material perpendicular to the force direction must be proportional to the magnitude of the force. In the case of bending or torsion, material must be appropriately distributed in the plane of the force and the area of cross-section must be proportional to the third power of the force. If a force is applied perpendicular to the longitudinal axis of a beam, bending occurs with compression on one side and tension on the other. In the center of the beam there is a neutral axis that curves without strain.

## Design of bone elements

To maximize resistance to bending stress there are two alternatives. The material can be entirely at the surface or it can be distributed in proportion to the length of the lever arm, with hardly any material in the center and increasingly more nearer to the surface.

In bone, the properties of the material itself can be adjusted, with more crystallization increasing the E-modulus on the side of the compression, and more fibers, increasing the tensile modulus on the other side. Another possible adjustment is found in the orientation and position of osteons and trabeculae (Dullemeijer and Fruitema 1982).

In human-made constructions, engineers distribute the material almost exclusively at the two surfaces, with a thin interconnecting bar between them. This bar serves only to keep the two surfaces apart and connected (using a T-bar or an I-bar). Alternatively, a tube is used which has the advantage that forces from various directions can be resisted.

The same factors are involved in animal constructions. Many vertebrates have more or less tubular bones, but there are so many exceptions that one is led to ask which mechanical principles are of critical importance here.

The common assumption that the cylindrical or tubular structure is cheapest needs to be put in perspective. Whether or not this statement is correct depends on the force pattern. If the bending force remains in one plane or varies only a little in direction, elliptical, narrow, rectangular, T-bar or I-bar designs require less material and so are cheaper than a cylinder. When we also recognize that the organism can influence the modulus of the material, we conclude that there must, almost invariably, be several good – maybe optimal – solutions to the same problem.

Thus bones can be constructed in various ways to accommodate the demands of varying mechanical requirements with different maximal solution.

## Conclusion

In the functional explanation of bony structures we have followed the principle of optimal design, that is, to construct a model for the shape and structure that is supposed to fulfill a demand with a minimal amount of material or costs. At various steps in this procedure we had to make choices or had to leave open which solution should be taken. Moreover, a quantitative prediction of volume and mass could not yet be made. Because of the inherent indeterminacy of the required construction, the organism had different choices, for example, standing on four legs where only two or three are needed.

Another important aspect is the implicit value of the safety factor. How much resistance against the forces is needed or, in other words, how much strain is allowed (and thus how much stress can be taken). The organism seems to have a perception mechanism of gradual changes in the loading so that it reconstructs the bony elements, making use of redundancy in statically indeterminate constructions or making use of its fast turnover of the building material.

Whether this kind of optimality contributes to an optimality in the sense of evolutionary theory remains to be proven. Therefore the significance of these structures to fitness must be determined. For the latter, *all* functions and necessary conditions should be considered; for example, next to giving mechanical support, the ability of bone of an easy, quick adaptability for growth and transformation for which a specific amount of strain in the bony elements must be kept as a necessary stimulus. It is ultimately the balance or compromise in the structure meeting various demands which determine the contribution to fitness.

## Further reading

Alexander, R. McN. (1982) *Options for Animals*. Arnold, London.
Alexander, R. McN. (1985) The ideal and the feasible physical constraints on evolution. *Biol. J. Linnean Soc.*, **26**, 345–58.
Barel, C. D. N. (1985) Concepts of an architectonic approach to transformation morphology. *Acta Biotheoretica*, **41**, 345–81.
Dullemeijer, P. (1974) *Concepts and Approaches in Animal Morphology*. Van Gorcum, Assen.

Dullemeijer, P. (1980) Functional morphology and evolutionary biology. *Acta Biotheoretica*, **29**, 151–250.

Dullemeijer, P. and Fruitema, F. (1982) The relation between osteon orientation and shear modulus. *Neth. J. Zool.*, **32**, 300–6.

Ramaekers, J. G. M. (1978) The rheological behaviour of cortical bone and cartilage of *Bos taurus*. *Acta Morphol. Neerl. Scand.*, **16**, 55–67.

Zweers, G. A. (1979) Explanation of structure by optimalization and systemization. *Neth. J. Zool.*, **29**, 418–40.

# 4.3 Optimization of musculoskeletal design – does symmorphosis apply?

ANDREW A. BIEWENER

As a proposed biological design principle, symmorphosis requires that the structural design of a system be matched to its functional requirements to ensure economy of design at multiple levels of organization. Implicit in the concept of optimization is that organisms are at equilibrium with their environment. Because of phylogenetic ancestry and selective forces posed by a constantly changing environment, this cannot strictly be the case. Hypotheses testing the adaptive significance of organismal design, however, usually assume an optimal relationship between structure and function. I argue here that, although the design of the musculoskeletal system cannot be considered to be optimized for any one broad functional role, such hypotheses can provide useful experimental tests of functional design. This section addresses the concept of symmorphosis by examining two aspects of musculoskeletal design in relation to the biomechanics of locomotion in mammals: (1) bone design in relation to strain distribution and skeletal remodeling; and (2) design of muscle–tendon systems for elastic energy savings versus control of movement.

### Bending, bone curvature, and load predictability

Bone is strongest when loaded in compression, being weaker in tension and weakest in shear. However, because limb bone elements are generally long, favoring increased stride length, they are susceptible to bending. Unless kinematic adjustments of limb movement are made to align bone elements more closely with externally (ground reaction) and internally (muscular) transmitted forces, increased bone length must lead to greater bending-induced stresses (or strains).

70

Larger animals, therefore, restrict their limb movements to a more para-sagittal plane and to smaller joint excursion ranges. These changes have the effect of aligning their limb bones more closely with the direction of ground reaction force throughout the period of limb support. While this favors a reduction of bending and greater axially directed compressive loading of the bone, available data indicate that most long bones experience significant stresses resulting from bending of the bone's shaft, in which 80–95 percent of midshaft strain (or stress) is the result of bending. While bending in smaller species may likely result from off-axis loading of the bone associated with their more "crouched" postures, bending in large species may actually be produced by longitudinal curvature of the bone itself, which induces a bending moment by axially directed compressive forces.

Why enhance bending? Symmorphosis, based on an optimality benefit–cost argument of strength versus mass, would argue the opposite. This indicates that one or more other factors must be considered. At least two possibilities for the role of bone curvature emerge. The first is simply a packaging requirement: "anti-gravity" or limb extensor muscles that support an animal's weight must be larger than those (limb flexors) functioning to swing the limb forward. Consequently, this requirement would argue that the posteriorly concave curvature of long bones provides a better mass distribution of the muscles that transmit force along the bone. A second possibility is that bone curvature provides an intrinsic control of bending direction, increasing the predictability that bending, and hence maximum stresses, will occur at a given location in the bone's cortex. The trade-off between load predictability and load carrying capacity can be modeled to fit the curvature observed in the radius of many species and is consistent with the longitudinal curvatures of other long bones.

While improved control of bending orientation costs additional bone mass to achieve a given strength, it has the advantage of providing a cross-sectional architecture that helps to assure an adequate safety factor. Although strength:mass is maximized with straight elements, this depends on the bone experiencing an axial load. If loading direction is allowed to vary, a straight bone with a minimal cross-sectional area might very well experience a catastrophic bending moment causing bone fracture. Curvature of the bone's shaft, although requiring greater bone mass to maintain stresses within safe limits, increases the predictability that stresses will not exceed the failure limit of the bone, potentially reducing fracture risk.

Clearly, a simple cost : benefit analysis of skeletal design based on compressive strength versus bone mass cannot explain the complexity of long bone design. The shortcomings of such an analysis, however, direct our attention to other factors previously unrecognized as being important. It also leads to a more integrative perspective of bone design in relation to functional loading, in which the composite organization of muscles relative to skeletal elements, and the forces that each must transmit, can be appreciated in the broader context of the kinematics and mechanics of limb support during locomotion.

### Muscle–tendon design: energy savings versus stiffness and control of movement

Although shortening to perform positive work is the most common view of muscle function, the need for muscles to stabilize motion or to dissipate energy is also important in locomotion and is often associated with the storage and recovery of elastic strain energy, most often in the muscle's tendon. The stiff, but high resiliency of collagen makes it an ideal material for functioning as a spring in the tendons of vertebrates: 91–95 percent of the energy stored upon stretching can be recovered through elastic recoil. Although muscles possess elasticity, the amount of strain energy that they can store is generally far less than that achieved by most tendons.

Using the concept of symmorphosis, we can define the optimal design ($U_{opt}$) of a muscle–tendon unit that functions to save energy by elastic storage in terms of the following benefit : cost ratio: $U_{opt} = E_{es}/(E_{metab} + E_{mass})$, where $E_{es}$ is the energy saved by elastic storage (tendon + muscle), $E_{metab}$ is the metabolic cost of force generation by the muscle, and $E_{mass}$ is the summed *relative* inertial cost of muscle mass versus tendon mass in the moving limb and their respective maintenance costs at rest. Given that the maintenance and operating cost per unit mass of tendon is less than that of muscle and that the mass of tendon needed to transmit a given force is also much less (300 to 500-fold) than that of muscle, this relationship is maximized when the required force is transmitted solely by the passive stretch of the tendon. The presence of a muscle costs an animal energy to operate, maintain and move relative to the mass of tissue that might otherwise be invested as tendon to transmit forces passively.

This functional distinction is schematized in Figure 4.1. Figure 4.1A shows a limb having a muscle in series with a tendon (depicted as a

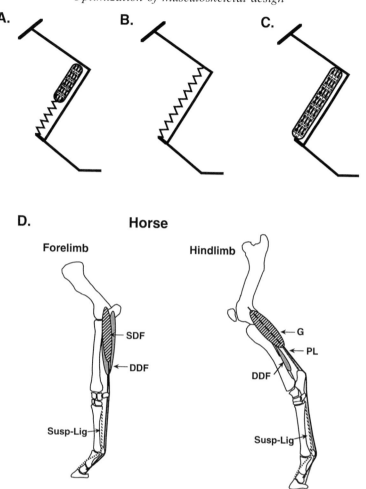

Figure 4.1. Limb models showing (A) muscle in series with a tendon (passive spring) reflecting a trade-off between control of joint excursion (muscle) and elastic savings (tendon); (B) passive elastic ligament maximizing elastic energy savings, ideal for a steady-state oscillatory system; and (C) muscle alone, with increased fiber length to maximize control of joint excursion. (D) Forelimb and hindlimb of a horse, illustrating the principal muscle–tendon units and ligaments that function as springs during gait (SDF, superficial digital flexor; DDF, deep digital flexor; Susp-Lig, suspensory ligament; P, plantaris; G, gastrocnemius).

spring). In this case, the muscle consumes energy to generate the forces necessary for strain energy storage in the tendon (as well as its own elastic elements); however, it provides for the possibility of controlling joint

excursion. Figure 4.1B shows a limb in which the joint moment is passively resisted by the tendon (spring) alone. This is the situation favored
for maximizing elastic energy recovery relative to the maintenance and
operating cost of tissue mass invested for energy savings. In this case,
however, joint excursion cannot be actively controlled. Figure 4.1C
shows a limb with a long fibered muscle but no tendon. This provides
for maximal control of joint excursion but limits elastic savings relative to
metabolic cost for force generation using long fibers and the inertial cost
of a larger muscle.

Are there examples in which elastic energy storage is optimized in the
limbs of animals? Ungulates show evidence of such an optimization in
three muscle–tendon systems; the most extreme being the long digital
flexor (or plantaris) muscle of the hindlimb (Figure 4.1D), in which the
muscle fibers are very short and greatly reduced. In horses, the reduction
of muscle fibers is extreme, indicating an evolutionary loss of muscle and
its replacement by collagen to form a long (0.85 m) ligamentous structure, spanning from the femur to the terminal phalanx. Similarly, the
superficial digital flexor muscle in the horse forelimb is also multipennate
with extremely short (5–7 mm) fibers that attach to a long (0.65 m) tendon, spanning from the radius to the intermediate phalanx. The extreme
length of the muscle's tendon makes it improbable that the fibers do little
more than generate force to provide for elastic stretch and recoil of the
tendon. Hence, the muscle fibers cannot provide useful control of carpal,
metacarpal or phalangeal joint excursion. Portions of the deep digital
flexor muscle and certain carpal flexors, with generally longer (35–
65 mm) fibers, presumably provide most of the control at these joints.
Horses also possess a third set of springs, the suspensory ligaments,
within their metacarpus and metatarsus, which also retain remnant
(interosseous) muscle fibers, indicating evolutionary loss of intrinsic
foot musculature to enhance elastic energy storage.

Another group of animals that exploit the use of elastic energy savings
are the kangaroos and wallabies. Elastic energy savings in these animals
are believed to explain their ability to hop at higher speeds with little, or
no, increase in metabolic energy expenditure. In a recent study, we found
that the stresses in the principal hindlimb tendons during steady-speed
hopping were nonuniform, being distributed among these tendons in a
manner that did *not* maximize elastic energy savings. Stresses were greatest in the shortest tendon (gastrocnemius), slightly less in the plantaris
tendon, and much lower in the digital flexor tendon (Figure 4.2A).
Consistent with their elastic energy savings function, the muscle fibers

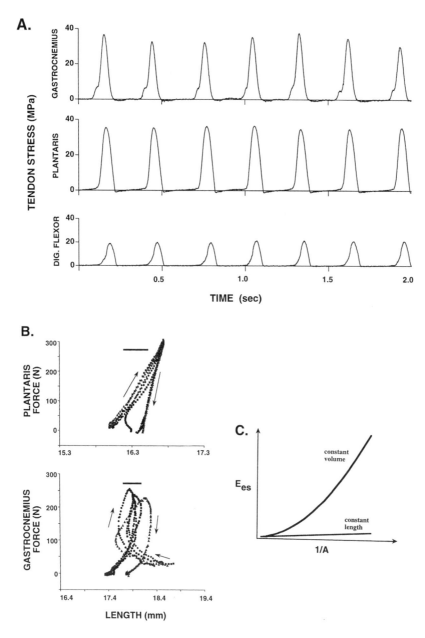

Figure 4.2.   (A) *In vivo* stresses acting in the plantaris, gastrocnemius, and digital flexor tendons of a tammar wallaby hopping at $4.5\,\mathrm{m\,s^{-1}}$. (B) *In vivo* force versus length of the plantaris and gastrocnemius during the stance phases of four successive hops (length change was measured directly by means of sonomicrometry). (C) Elastic energy savings ($E_{es}$) versus the inverse of tendon area ($1/A$). With length constant, $E_{es}$ increases linearly; with volume constant, $E_{es}$ increases exponentially ($\alpha 1/A^2$).

of gastrocnemius and plantaris operate like a stiff spring, developing force with little change in length ($<2$ percent; Figure 4.2B). However, despite the general reduction of muscle volume and fiber length versus tendon, maximal strains were not developed in the longest tendons (plantaris and digital flexor) which would have favored greater elastic storage (Figure 4.2C).

Why do wallaby tendons diverge from these expectations for maximizing elastic energy savings? The observation that many vertebrate tendons operate with very high safety factors ($\geq 6$–8) has been made by several investigators, who have suggested that maintaining adequate stiffness, in series with the muscle's fibers, is a competing requirement of many tendons for muscular control of displacement and joint excursion. In the case of the wallaby, the low stresses acting in the digital flexor tendon compared with the gastrocnemius and plantaris tendons indicate the importance of this tendon in controlling digital extension during the hop. Steady-speed hopping may be the exception for wallabies much of the time. As for other species, accelerative movements, changing direction, and establishing a stable base of support with the foot make control of foot position a key functional requirement of muscle–tendon design that precludes optimization for elastic savings.

In summary, examples of evolutionary optimization for elastic savings in certain muscle–tendon units of terrestrial mammals can be found. Extreme designs predicted by optimality arguments, in which muscle fiber length and volume are reduced to the extent that passive ligamentous springs replace muscles and their tendons, however, are rare. They are rare owing to at least two constraints: (1) the historical constraint of phylogeny; and (2) the need to maintain adequate stiffness for control of movement. The range of muscle–tendon architectures observed in a diversity of animals suggests that economy versus control of movement interact in a complex fashion as competing demands that are often not easily described in simple cost–benefit terms.

### Further reading

Alexander, R. M. (1981) Factors of safety in the structure of animals. *Sci. Prog.*, **67**, 119–40.

Alexander, R. M. (1988) *Elastic Mechanisms in Animal Movement.* Cambridge University Press, Cambridge.

Bertram, J. E. A. and Biewener, A. A. (1988) Bone curvature: sacrificing strength for load predictability? *J. theor. Biol.*, **131**, 75–92.

Biewener, A. A. (1990) Biomechanics of mammalian terrestrial locomotion. *Science*, **250**, 1097–103.

Currey, J. D. (1984) *The Mechanical Adaptations of Bone*. Princeton University Press, Princeton, NJ.

Ker, R. F., Alexander, R. M. and Bennett, M. B. (1988) Why are mammalian tendons so thick? *J. Zool., Lond.*, **216**, 309–24.

Lanyon, L. E. (1987) Functional strain in bone tissue as an objective, and controlling stimulus for adaptive bone remodelling. *J. Biomech.*, **20**, 1083–93.

# 4.4 Responses of bone to stress: constraints on symmorphosis

DANIEL E. LIEBERMAN and
ALFRED W. CROMPTON

Symmorphosis, the hypothesis that organisms can adapt "economically" to changing functional demands, is not new to bone biology. Meyer first applied an early version of the theory to trabecular bone in 1867, which was later incorporated into Wolff's 1892 Law of Bone Transformation that every change in a bone's function is followed by changes in its internal structure and external conformation. Despite widespread evidence that bones respond dynamically to external stimuli, Wolff's Law remains controversial because the processes by which they adjust to their mechanical environment are poorly understood.

We discuss here, using experimental data, several major processes by which bones respond to mechanical loading and how these processes constrain the optimization of mechanical design. Stress ($\sigma$, $F/A$) generates strain ($\varepsilon$, $\Delta L/L$) that can induce osteogenesis because it potentially damages bone tissue. Frequent, high-strain magnitudes often induce bone growth, whereas low levels lead to resorption. Professional tennis players, for example, develop extreme hypertrophy in their playing arm, while astronauts in the gravity-free environment of space resorb bone systemically. Because bone is a complex tissue with many functions and structures, different types of bone (for example, woven, parallel-fibered, or lamellar) often respond to similar strains differently. For example, chewing elicits comparable strains in the mandibular corpus of the goat and the opossum which have different histologies, but, as we discuss below, only the goat responds with a significant amount of Haversian remodeling, even though remodeling can occur elsewhere in the opossum jaw.

## A hierarchical model for bone symmorphosis

A simple model expresses how bones may adapt to mechanical loading:

Structural responses/constraints

Force → Transductional responses/constraints → Adaptation (modeling/remodeling)

Design constraints

Force applied to a bone, measured as stress, generates strain that can vary in magnitude, frequency, predictability, and orientation. Such stimuli can ultimately lead to diverse adaptive responses. Macrostructurally, bones adapt either through *modeling*, the deposition of new bone, or *remodeling*, the resorption and/or replacement of old bone. Bones also adapt microstructurally in terms of collagen organization, mineralization, and so forth.

The intermediate processes and constraints that mediate osteogenic responses to force occur through at least three hierarchical levels: structural, transductional, and design. The most basic level of response is *structural*, the attributes of a bone that determine how a given force elicits strain. These include its macrostructure (size and shape), as well as many microstructural properties (for example, collagen orientation and density, mineralization, lamellar organization). Fibrolamellar bone, for example, has a higher tensile strength than Haversian bone, and young bone is more elastic than old bone. The second, *transductional* level comprises the cellular processes by which bone cells detect and respond to strain. These constraints are incompletely understood, but several factors appear to be critical. Bones may sense strain through nerves in the periosteum, and perhaps through pressure changes in fluid-filled canaliculi of osteocytes and from piezoelectrical potentials generated by collagen deformation. Vascular supply also limits bone responses to force because bone cells require nutrients. Finally, numerous design constraints operate at different levels involving numerous cell types. Some bones conserve mass, grow to specific dimensions, or have special structural properties. Most bones also perform non-mechanical functions that influence their shape.

Force, in other words, induces modeling and remodeling responses in bone through a series of interrelated, hierarchical processes.

## Variations in responses to strain

To examine how the structural, histological, and design levels of osteogenic response mediate the adaptation of bones to environmental

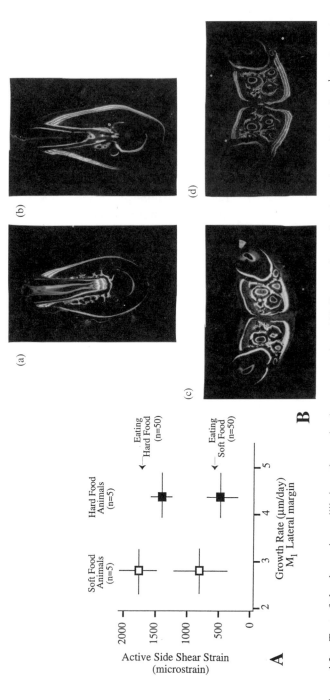

Figure 4.3. Test of the dynamic equilibrium theory in the opossum jaw. (A) Comparison of periosteal growth (μm day$^{-1}$) at lateral margin of corpus at $M_1$ with mean active side shear strains (μɛ) measured at the lateral margin of corpus near ventral margin at $M_1$. Strains in animals raised on hard food for 12 weeks ($n = 5$) are significantly lower ($p < 0.05$) than those in animals ($n = 5$) raised on soft food because of their faster growth rate ($p < 0.05$). (B) Comparison of growth in mandibular corpus for representative hard food (h) and soft food (s) animal at $M_1$ (M) and in symphysis (S). Fluorescent dyes (calcein and tetracycline) show that periosteal growth was significantly greater ($p < 0.05$) in the medial and lateral margins of the corpus and in the ventral margin of the symphysis. Note that the only Haversian systems occur along the medial margins of the symphyseal plate.

conditions, we present data from several experiments in which we subjected diverse species to reasonable ("normal") levels of habitual force in the laboratory. These illustrate a wide range of responses to increased levels of force.

### Macrostructural responses

If bone tissue is designed to tolerate certain strains, then bones should respond to increased forces structurally by augmenting their mass in their principal planes of deformation, thereby decreasing the level of strain a given force generates. This theory of dynamic strain similarity – formulated by researchers such as Lanyon, Rubin, and Biewener – explains many of the quantitative increases in bone mass at functionally equivalent sites in growing animals. To examine how this optimization process operates on mature animals we divided ten twelve-month-old female opossums (*Didelphis virginia*) into two equal groups that ate nutritionally identical hard and soft food, respectively, for twelve weeks. We labeled newly deposited bone with fluorescent dyes, and placed rosette strain gauges on the lateral surface of each mandible.

The dynamic strain similarity theory predicts that the opossums fed a soft-food diet would experience higher strains when chewing hard food than the opossums raised on hard food. The results (Figure 4.3) agree with the predictions. During the experiment, the hard-food animals deposited periosteal bone more rapidly in their mandibles (Figure 4.3B), particularly on the lateral surface, so that they experienced roughly 25 percent and 68 percent lower shear strains at this location when chewing hard and soft food, respectively (Figure 4.3A). In other words, the opossums adapting to a hard-food diet had lower strains than those who ate soft food because of osteogenic responses to the higher strains they previously encountered.

### Microstructural responses

Strains also induce microstructural changes, the most important of which is Haversian remodeling. Several factors, including strain magnitude and frequency, tissue age, and the degree of vascularization, appear to influence Haversian remodeling.

We examined how some of these factors affect Haversian remodeling by comparing mandibular growth in three groups of subadult goats (*Capra hircus*) fed different diets: all ate a hard, nutritionally adequate

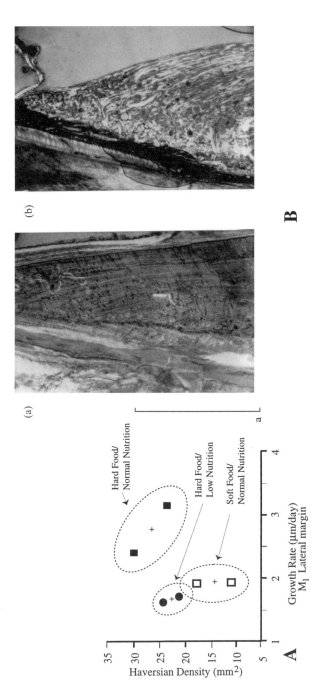

Figure 4.4. Relationship between strain, growth rate, and Haversian remodeling. (A) Relationship between growth rate (μm/day) and Haversian density (mm²) at the lateral margin of mandibular corpus at $M_1$ in three groups of goats. The hard food/normal nutrition goats had a significantly faster rate of growth and more dense Haversian bone than either the low nutrition group or the soft food group ($p < 0.05$). (B) Comparison of lateral margin of $M_1$ (a) goat and (b) opossum, raised on hard food. Note that goat has Haversian remodeling of fibrolamellar bone, whereas opossum has parallel-fibered bone without Haversian systems.

diet during the first and last four months of the experiment; during the middle four months, however, one group ate softened food, and one group ate a hard but nutritionally deficient diet (34 percent lower in protein and 50 percent lower in minerals). Fluorescent dyes were used to label bone growth and remodeling, which Figure 4.4A plots as the periosteal growth rate along the lingual margin of the mandible at $M_1$ versus Haversian remodeling density in the same region. The goats fed the hard, control diet had a higher rate of periosteal growth than the other goats ($p < 0.05$); and all the goats fed hard food, regardless of nutrient content, had more Haversian remodeling than those who ate soft food ($p < 0.05$). The goats whose jaws grew more slowly from less nutrition remodeled at a higher rate than those whose jaws grew more slowly from a lower strain regime.

These results confirm previous claims by Bouvier, Hylander, and others that strain levels influence both the rate of modeling and the rate of Haversian remodeling. Interestingly, goats' jaws grew significantly ($p < 0.05$) slower ($2.1 \, \mu\text{m day}^{-1}$) than the above-discussed opossums ($3.5 \, \mu\text{m day}^{-1}$), which were at a developmentally equivalent stage. These rate differences may be reflected histologically (Figure 4.4B). The periosteal bone deposited in the goat mandibular corpus is entirely fibro-lamellar, which is more organized and vascular than the parallel-fibered bone deposited in the opossum corpus. Peak strains in the goats and opossums were very similar, however. Peak shear strains on the active side at $M_1$ was $442.3 \pm 129.2 \, \mu\varepsilon$ ($n = 10$) and $484.4 \pm 275.9 \, \mu\varepsilon$ ($n = 50$) in the hard-food goats and opossums, respectively.

### Design constraints

Bones may adapt diversely to strain through different proportions of modeling and remodeling in order to select for certain properties such as shape, elasticity, density, or mass. Such design constraints occur to some extent through modulations of various cellular responses that regulate modeling and remodeling.

Varying responses to strain can be tested in limbs. In most vertebrates, distal limb bones tend to be thinner and, therefore, weaker than more proximal bones. While a purely structural model for bone optimization predicts that animals will compensate by depositing relatively more bone in their distal than proximal limb elements, natural selection may also favor animals who minimize mass distally to conserve the angular momentum of the limb during deceleration and acceleration (big feet take exponentially more kinetic energy to swing than small feet). Haversian remodeling rates are therefore predicted to increase in proportion to the square of each element's radius of gyration (an approximation of its moment arm) in order to repair the presumably higher rate of damage potentially generated by high forces in more lightly built bones.

We tested the relationships between moment arm length, modeling and remodeling in the limbs of six miniature swine (*Sus scrofa*) between the ages of one and four months. These animals were divided in two groups, of which one ran on a treadmill at $4.8 \, \text{km h}^{-1}$ twice daily for a total of 60 min. Bone growth was labeled with fluorescent dyes. As predicted above, the runners had significantly more rapid modeling rates at bone midshafts than the controls, but responses to strain differed between elements, suggesting that they have different design constraints. In particular, relative rates of midshaft modeling and remodeling varied inversely between proximal and distal elements in proportion to estimated moment arm length. In the hindlimb, for example, the pigs had a 71-fold higher rate of remodeling in the metatarsals than femur, but a concomitant 5.6-fold decrease in midshaft strength. This trade-off between modeling and remodeling is shown in Figure 4.5, which graphs the percentage of cortical bone modeled and remodeled relative to midshaft cross-sectional area for each element. Distal bones in the limb are weaker than the proximal bones to conserve mass, but appear to maintain this mechanical compromise with higher Haversian remodeling rates. Note that Haversian density is significantly greater for all bones (except the femur, which did not remodel) in runners than in controls, and that the modeling–remodeling trade-off slope differs between the forelimb and hindlimb, presumably because more of a pig's mass is concentrated cranially.

Bones respond to increased mechanical force through diverse optimization processes. However, any simple, direct relationship between exogenously induced strain and endogenous structural and microstructural responses remains elusive.

## Discussion

Modeling and remodeling processes coadjust to adapt to force-induced strain as predicted by the theory of symmorphosis, but these responses are varied and sometimes unpredictable. As we have shown, bones sometimes respond purely macrostructurally by altering their shape and size because a given force generates less strain in a more massive bone. In

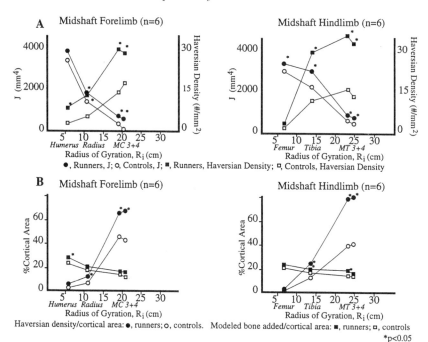

Figure 4.5.   (A) Polar moment area, $J$ (mm$^4$), and density of Haversian remodeling (mm$^2$) in midshaft of limb bones in pigs versus the estimated radius of gyration (moment arm length). Modeling and remodeling rates are higher in exercised animals; in both groups the distal limb bones are increasingly thinner and weaker, but appear to compensate with a correspondingly higher density of Haversian systems. (B) Evidence for a trade-off between modeling and remodeling; the percentage of modeled versus remodeled bone per cortical area at limb midshafts versus the estimated radius of gyration. Distal elements apparently conserve mass by modeling less and remodeling more; this trade-off is exaggerated in exercised animals.

other bones, intermediate mechanisms appear to preferentially induce Haversian remodeling instead of modeling. Most bones probably adapt to mechanical forces from both processes; in some such cases, the relative degree to which bones change through macrostructural and microstructural responses appears to vary inversely according to design constraints. This trade-off between modeling and remodeling suggests that the primary role of Haversian remodeling is not to recover mineral from bone tissue but instead to repair the fatigue damage induced by strain.

Symmorphosis, therefore, appears to be a real phenomenon in bones; one that we should expect, given the unpredictability of the interactions between an organism and its environment. Since symmorphosis is essentially the result of several hierarchical levels of process (structural responses, transduction, and design constraints), it follows that the history and nature of these processes constrain how bones optimize their mechanical design. Secondary remodeling, for example, requires vascularization, which, in turn, is a consequence to some extent of its rate of growth. Haversian remodeling, therefore, may be less common in active but more slowly growing reptiles than in equally active but more rapidly growing mammals because their bones are less vascularized.

The many functions of bone and the many processes by which it grows, however, confound efforts to construct simple models for the optimization of bone to mechanical demands. Future efforts to understand the adaptations of bones need to focus on the many poorly understood cellular processes that intervene between the proximate action of force on a bone and the ultimate activation of bone cells. More information on these processes will undoubtedly lead to a more satisfying but probably more complicated model of symmorphosis in bone.

### Further reading

Biewener, A. A., Swartz, S. M. and Bertram J. E. A. (1986) Bone modeling during growth: dynamic strain equilibrium in the chick tibiotarsus. *Calcified Tissue International*, **39**, 390–5.

Bouvier, M. and Hylander, W. L. (1981) Effect of bone strain on cortical bone structure in macaques (*Macaca mulatta*). *Journal of Morphology*, **167**, 1–12.

Currey, J. (1984) *The Mechanical Adaptations of Bones*. Princeton University Press, Princeton.

Frost, H. M. (1986) *The Intermediary Organization of the Skeleton*. CRC Press, Boca Raton.

Martin, R. B. and Burr, D. B. (1989) *Structure, Function, and Adaptation of Compact Bone*. Raven Press, New York.

Rubin, C. T. and Lanyon, L. E. (1984) Dynamic strain similarity in vertebrates: an alternative to allometric limb scaling. *Journal of Theoretical Biology*, **107**, 321–7.

# 5

# Muscles and locomotion

## 5.1 Overview

JAMES W. GLASHEEN

Although muscle plays a crucial role in digestive, reproductive, and circulatory systems, it is when we explore its role in locomotion that we begin to understand the relationship between structure and function. The functional results of the design constraints of muscle are seen at all levels of organization; ranging from the molecular interactions between actin and myosin to whole-animal performance.

On the molecular level, the cyclic interactions between a myosin head and an actin filament allow a muscle fiber to produce force and change length. To a large degree, the attachment rates of the myosin head to the actin filament determine the rate of force generation, while detachment rates determine maximum shortening velocity. Interestingly, though, when one measures the activity of enzymes associated with myosin-head *attachment*, it is correlated with maximum shortening velocity, and therefore myosin head *detachment*. The outcome of this tight correlation is that in spite of a broad variation in maximum shortening velocities, isometric muscle stress for all skeletal muscle is constant. This result proves especially interesting when one considers that if a fast-twitch, anaerobic muscle is exposed to a chronic stimulation pattern, the myosin heavy-chain isoforms will slowly be replaced by increasingly slow-twitch, aerobic isoforms. Thus, the transformed muscle will have a vastly different shortening velocity, but will have the same isometric muscle stress as before the transformation.

Next, let us explore how the muscle functions within the animal. Muscles can operate over a range of sarcomere lengths and shortening velocities, but is the manner in which the muscle is used matched to its functional demands? The demands on a locomotor muscle are determined, in large part, by the medium in which the animal moves. A fish swimming through the water must do work on the environment as it accelerates fluid rearward. Thus, in addition to being operated at moderate lengths and shortening velocities, the locomotor muscles of a fish must be stimulated at an appropriate phase in the length cycle to produce power efficiently. In contrast, terrestrial runners do relatively little work on the environment. Thus, they have the opportunity to store energy during one phase of a step and recover that energy during a subsequent phase. For example, the locomotor muscles of running turkeys are attached to elastic tendons which store and release energy in each step. Because the tendons do much of the work of each step, the muscles need to shorten relatively little and are therefore able to produce force economically.

Finally, with an understanding of muscle structure and function, we can explore whole-animal performance. We can develop models which predict optimal locomotor strategies for swimming fish which address the question: At what speed should a fish swim to travel a maximum distance per unit energy consumed? When we compare our theoretical predictions of optimal speed for migrating fish with data on actual swimming speeds, we see that some fish adopt locomotor strategies which allow them to minimize the metabolic cost of locomotion. However, equally interesting are the instances in which the animals' behaviors do not match the predictions of the model. These instances force us to probe not only the *functional* constraints to which animals are exposed, but also the *historical* and *developmental* factors which shape an animal's structure.

# 5.2 The malleability of skeletal muscle

DIRK PETTE and ROBERT S. STARON

Skeletal muscles are composed of a wide variety of fiber types, each with a specific protein isoform pattern and metabolic enzyme activity profile. Among the factors that determine the fiber type composition of a muscle, neuromuscular activity plays an important role. Differences in fiber type composition may, therefore, be regarded as adjustments to specific functional demands. However, intrinsic programs and species-specific properties are also important in determining the fundamental patterns of expression.

## The multiplicity of muscle fiber types

Distinct muscle fiber types differ in their myosin complement. Myosin is a hexameric protein consisting of two heavy chains (HCs) and four light chains (LCs). The HC portion, expressed as a member of a multigene family, can be detected using histochemical, immunohistochemical, or biochemical methods and forms the basis for fiber type delineation. To date, at least 10 myosin HC isoforms have been identified in adult extrafusal fibers. Although some of these isoforms are apparently restricted to a small number of specific muscles, theoretically at least ten different HC-based fiber types can be identified. However, the fiber populations of most skeletal muscles in small mammals seem to encompass only three fast fiber types and one slow fiber type. The very fast type IIB fibers express myosin HCIIb; the slightly less fast type IID/X fibers, HCIId/x; and the slowest of the fast group, type IIA fibers, the HCIIa isoform. Type I fibers express HCI$\beta$, thought to be identical with the cardiac $\beta$-myosin HC. It should be noted, however, that recent evidence points to the existence of an additional slow fiber type expressing an $\alpha$-cardiac-like HC isoform.

89

In addition to these "pure" fiber types which express only a single HC isoform, skeletal muscles also contain "hybrid" or transient fibers which express two (and occasionally more) myosin HC isoforms in varying ratios. Muscle fiber diversity is further increased by various combinations of other protein isoforms. For example, different combinations of myosin LC isoforms affect the contractile properties within a defined HC-based fiber population. Similarly, one finds a relatively unexpected metabolic heterogeneity, both within and between the major fiber populations. Taken together, these qualitative and quantitative differences in gene expression result in a large spectrum of gradually differing muscle fibers forming a continuum of types between the very fast, low oxidative and very slow, high oxidative.

### Plasticity of muscle fiber phenotype

Muscle fibers are not fixed elements but represent versatile entities, capable of changing their phenotype in response to altered conditions (Figure 5.1). Innervation, total amount of contractile activity, and specific hormones represent major factors significantly affecting the phenotypic properties of adult muscle fibers. By increasing neuromuscular activity (via, for example, exercise or chronic electrostimulation), overloading a muscle (via, for example, stretching or surgical ablation of synergists), or decreasing levels of thyroid hormone, a fast muscle can be made slower. Conversely, by decreasing neuromuscular activity (via, for example, reduced contractile activity, detraining, denervation, spinal cord transection), unloading (via, for example, immobilization, hindlimb suspension, microgravity), or increasing levels of thyroid hormone, a slow muscle can be made faster. This malleability of skeletal muscle is

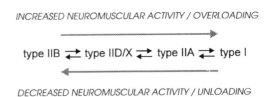

Figure 5.1. Schematic representation of reversible fiber type transitions in muscles of small mammals in response to altered functional demands. Increased neuromuscular activity shifts fiber types in the direction from fast to slow, whereas transitions in the opposite direction occur under conditions of reduced neuromuscular activity.

perhaps best illustrated by experiments where contractile activity of fast muscles is artificially increased by delivering, via implanted electrodes, a low-frequency impulse pattern normally transmitted to slow muscles. As compared to other models that increase neuromuscular activity, chronic low-frequency stimulation delivers a standardized and reproducible regimen of contractile activity. Also, because this form of increased activity maximally activates all motor units of the target muscle, it elicits a full range of adaptive responses. Finally, chronic stimulation provides the possibility of establishing dose–response relationships by altering the workload and following the time course of the induced changes.

Chronic low-frequency stimulation of fast muscles elicits a series of orchestrated changes which affect all functional elements of the muscle fiber. As a result, a fast-twitch, fatigable muscle is gradually transformed into a slower contracting, less fatigable muscle with lower force output. This fast-to-slow transformation induces profound changes in the expression of numerous genes. Thus, chronic low-frequency stimulation modifies the isoform pattern of contractile and regulatory myofibrillar proteins, the protein composition of the $Ca^{2+}$-regulatory system, and the enzyme activity profile of energy metabolism. Briefly, the fast-type isoforms of myosin heavy and light chains, the troponin subunits, and the $Ca^{2+}$-ATPase of the sarcoplasmic reticulum are exchanged in a time-dependent manner with their slower counterparts. In addition, the expression of parvalbumin, a cytoplasmic $Ca^{2+}$-binding protein specific to fast fibers, is suppressed, whereas the expression of phospholamban, a regulatory protein of the sarcoplasmic reticulum $Ca^{2+}$-ATPase, normally present only in slow fibers and in cardiac muscle, is upregulated.

At the cellular level, the transformation process corresponds to sequential transitions of fiber types in the direction from fast to less fast and from less fast to slow (Figure 5.1). This explains why transforming muscles display an elevated fraction of hybrid fibers. Corresponding to the sequential transitions in myosin HC isoform expression, the spectra of coexisting isoforms in these fibers change with the duration of stimulation. The next-to-the-last step of the fast-to-slow conversion is characterized by a pronounced increase in fibers displaying a fast/slow HC combination (HCIIa/HCIß). Because fast- and slow-type myosin light chains are capable of combining with fast and slow myosin HC isoforms to form various isomyosins, the time-dependent changes in light chain composition do not strictly parallel the alterations of the HC isoform pattern. As a result, a multitude of hybrid isomyosin molecules can be expected to assemble during the transformation process.

Among the changes induced by increased activity, metabolic adaptation is of special importance. Sustained contractile activity imposed upon fast fibers metabolically programmed for short-term, phasic activity greatly enhances the demand for energy supply. However, this cannot be satisfied by a predominantly glycogen-based metabolism. Therefore, elevated capacities for glucose transport and glucose phosphorylation are among the first adaptive responses. Moreover, there are marked increases in the activity levels of enzymes involved in aerobic-oxidative pathways, such as fatty acid oxidation, ketone body oxidation, the citric acid cycle, the respiratory chain, and the phosphocreatine shuttle. The increases in enzyme activities of terminal substrate oxidation occur in parallel with pronounced elevations in mitochondrial volume, reaching levels which greatly exceed those found in normal slow muscles. Increases in capillary density concomitant with decreases in fiber cross-sectional area serve to enhance oxygen supply. Taken together, these and other changes at the cellular and molecular level reflect a profound rearrangement of energy metabolism with a shift from anaerobic to aerobic metabolic pathways.

### Species-specific ranges of adaptation

Although the effects of chronic low-frequency stimulation have been studied in various mammalian species (including mice, rats, guinea pigs, rabbits, cats, dogs, goats, sheep, and humans), the results are not easy to compare because of differences in stimulation protocols, duration of stimulation, and so on. Nevertheless, in most cases it has been shown that low-frequency stimulation induces fast-to-slow transitions and enhances the aerobic-oxidative capacity of energy metabolism. However, species differences exist and fast-to-slow transitions are easier to induce in larger than in small animals. For example, low-frequency stimulation of mouse fast-twitch muscles does not lead to noticeable changes in the pattern of myosin HC isoforms. Likewise in the rat, stimulation-induced fiber type transitions mainly occur within the fast fiber population (type IIB $\rightarrow$ type IIDX $\rightarrow$ type IIA) and the ultimate transitions from fast to slow (type IIA $\rightarrow$ type I) are difficult to achieve. This already narrow range of stimulation-induced transition is even more restricted when combined with hyperthyroidism, causing only moderate transitions within the fast fiber subtypes or myosin HC isoforms. Conversely, combining hypothyroidism in the rat with chronic low-frequency stimulation results in markedly enhanced fast-to-slow fiber type conversions.

Species-specific ranges of adaptation also exist with regard to activities of key enzymes of energy supply. Chronic low-frequency stimulation causes small increases in mitochondrial enzyme activities in the mouse and large increases in the rabbit, with intermediate changes in the rat and guinea pig. Interestingly, the increases in enzyme activities appear to be inversely proportional to their basal levels in unstimulated muscles. These findings suggest that fast muscles of small mammals are metabolically programmed for sustained activity and, therefore, cannot be induced to attain higher expression levels of their enzyme apparatus for aerobic-oxidative metabolism.

## Conclusions

Skeletal muscle consists of a multitude of fiber types and, therefore, is an extremely heterogeneous tissue. Different fiber types arise by various combinations of regulatory and contractile protein isoforms coupled with a continuum of metabolic enzyme activity profiles. A fundamental characteristic of muscle fibers is their capability to respond to altered functional demands by qualitative changes in their expression patterns of myofibrillar and regulatory protein isoforms and quantitative adjustments in their metabolic enzyme profiles. Adaptive responses to long-lasting changes in functional demands, therefore, consist of a remodeling of muscle fiber phenotypes. Depending on whether exposed to increased or decreased functional demands, changes in gene expression result in gradational fiber type transitions from either fast-to-slow or slow-to-fast. At the molecular level, coordinate changes in the isoform patterns of various myofibrillar and other proteins point to preferential combinations of distinct functional elements, corresponding to "fiber type-specific modules." These modules result from synchronized events of gene upregulation and downregulation. The underlying regulatory mechanisms are unknown, but obviously apply to different levels of organization. For example, the expression of different myosin HC isoforms relates to the control of different members of a multigene family, whereas other protein isoforms (for example, troponin-T) are generated by alternative splicing.

The finding that the fast-to-slow fiber transformation involves sequential and parallel changes in the expression of several myofibrillar and other proteins points to the existence of thresholds. These allow graded responses to different amounts of workload. The nature of these hypothetical thresholds is obscure, but it has been speculated that a reduction in energy potential of the muscle fiber might play an important role.

In support of this notion, the phosphorylation potential of the ATP system is markedly depressed in muscles exposed to chronic low-frequency stimulation.

Alterations in myosin composition have important energetic implications. The exchange of fast, high ATPase activity myosin with isoforms of lower ATPase activity causes not only a decrease in the velocity of shortening, but creates conditions for more economical use of energy for contractile activity. Similarly, the metabolic changes concomitant with the fast-to-slow transformation in low-frequency stimulated muscles can be interpreted as compensation for an energetic imbalance. Thus, the switch to aerobic-oxidative pathways with higher ATP yields helps improve the energy supply for enhanced and sustained contractile activity.

In summary, the malleability of skeletal muscle provides impressive examples in support of the symmorphosis concept. Fiber type transitions occurring under the influence of increased or decreased workload demonstrate the dynamic state in which muscle fibers exist. Thus, the observable changes in molecular, cellular and functional properties characteristic of fiber type transitions emphasize the notion that a distinct fiber type represents a structurally and functionally designed unit for a specific type of contractile activity.

### Further reading

Bottinelli, R., Canepari, M., Reggiani, C. and Stienen, G. J. M. (1994) Myofibrillar ATPase activity during isometric contraction and isomyosin composition in rat single skinned muscle fibers. *J. Physiol. (London)*, **481**, 663–75.

Crow, M. M. and Kushmerick, M. J. (1982) Chemical energetics of slow- and fast-twitch muscles of the mouse. *J. Gen. Physiol.*, **79**, 147–66.

Pette, D. and Staron, R. S. (1993) The molecular diversity of mammalian muscle fibers. *News Physiol. Sci.*, **8**, 153–7.

Pette, D. and Staron, R. S. (1997) Mammalian skeletal muscle fiber type transitions. *Int. Rev. Cytol.*, **170**, 143–223.

Pette, D. and Vrbová, G. (1992) Adaptation of mammalian skeletal muscle fibers to chronic electrical stimulation. *Rev. Physiol. Biochem. Pharmacol.*, **120**, 116–202.

Schiaffino, S. and Reggiani, C. (1996) Molecular diversity of myofibrillar proteins: gene regulation and functional significance. *Physiol. Rev.*, **76**, 371–423.

# 5.3 Fine tuning the molecular motor of muscle

## H. LEE SWEENEY

In moving the skeleton, pumping blood and controlling movements in the viscera and vasculature, muscle performs mechanical work in animals. Given the large muscle mass of an animal, it is generally considered that the design of muscle must be optimized to perform its function with minimal energy cost and maximal efficiency. In order for this to occur, a number of processes within muscle must be altered in parallel as the functional demand on the muscle is changed. Of central importance is the design of the molecular motor, myosin, which – via its cyclic, ATP-consuming interaction with actin – accounts for most of the energetic cost of muscle contraction.

One might expect that given the differences in functional demands placed on the three types of mammalian muscle (skeletal, cardiac, and smooth), fundamental differences should exist in the myosin motor isozymes expressed in these tissues if optimization has occurred. Smooth muscle cells must generate forces for prolonged periods of time with little shortening. Thus a myosin capable of generating high isometric force, with low energy cost (high economy) would be optimal. Skeletal muscle is heterogeneous in its myosin isozymes and in the functional demands that it faces. Thus the myosin isozymes of skeletal muscle must be based on compromises that provide for high economy in muscles that are used in a predominately isometric pattern, and myosin isozymes that can provide high shortening velocities and power outputs with maximal efficiency in muscles that perform shortening work. For cardiac muscle, the demands, and thus the optimal solution, should be similar to that for skeletal muscle in that both economical isometric contractions and efficient, but powerful, shortening contractions must be generated.

### Constraints on the myosin motor

Ultimately, the design of any myosin motor involves compromises in the amount of force generated per ATP consumed, the velocity of movement, and in the rate at which ATP is consumed. These compromises arise because constraints are placed on the myosin motor by the architecture of muscle cells, in which myosin is organized into filaments, which ultimately are arranged in repeating units. This arrangement is necessary in order to place myosin motors in parallel so that the piconewton level forces of single myosin molecules can be added together to produce substantial macroscopic forces; and to place myosin motors in series so that the nanometer excursions of single myosin molecules can be added together to produce macroscopic movements. Irrespective of the motor design, the length of the filaments themselves represents a compromise between the attainable force and shortening velocity.

To understand the nature of the constraint that this filament organization imposes on the design of a myosin motor, one must first recognize that in order for molecular motors to transmit force and movement, a cyclic interaction with a cytoskeletal structure (actin filaments in the case of myosin motors; tubulin microtubules in the case of kinesin and dynein motors) must occur. This motor cycle (referred to as the cross-bridge cycle in the case of myosin) minimally consists of an attached state (that is, bound to the force-transmitting filament) and a detached state, both of which may be further subdivided into additional states. In going from detached to attached state(s) and back to detached, the motor goes through one complete cycle, converts one molecule of ATP to ADP and inorganic phosphate, and produces force and movement.

If one considers a single motor molecule working in isolation, then the optimal design would be for the fraction of the cycle during which the motor was attached and producing force to be much greater than the fraction of the cycle during which the motor is detached and thus not transmitting force (that is, a duty cycle approaching one). This design would optimize the amount of force generated per cycle. While this design is realized for the molecular motor kinesin, which itself is not organized in filaments and is involved in intracellular movements, this design is inappropriate for filamentous myosin. Myosin motors operate in large ensembles and asynchronously in filaments. These filaments are in turn bundled into arrays of filaments, which in striated (skeletal and cardiac) muscle cells are organized into repeating structures known as sarcomeres (Figure 5.2). Since the actions of individual myosin motors are asynchronous, the duration of the attachment part of the cycle ultimately limits the shortening velocity, because attached myosin cross-bridges create a drag. Experiments have demonstrated that as the velocity of shortening of a muscle increases, the number of attached myosin cross-bridges decreases until the maximal velocity is reached when the myosin cross-bridges promoting shortening are counter-balanced by attached cross-bridges that are opposing shortening. Thus the maximal velocity is ultimately limited by the rate at which attached myosin cross-bridges can detach. From the standpoint of maximizing shortening velocity, the time spent in attached (force producing) state(s) should be small, but from the standpoint of maximizing force per ATP utilized, the time spent in these state(s) should be large.

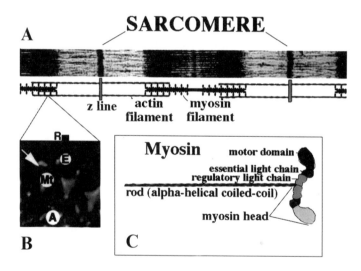

Figure 5.2. (A) Filament structure of the sarcomere of striated muscle as revealed by a section of an electron micrograph (above) of a rabbit muscle. Below the micrograph is a schematic diagram depicting the z line (site of actin filament attachment), actin (thin) filament, and myosin (thick) filament. (B) The actin–myosin interface as revealed by cryo-electronmicroscopy. The actin filament is labeled, A, while the parts of the myosin head are labeled Mt (motor domain), E (essential light chain), and R (regulatory light chain). (C) Schematic representation of a single myosin molecule.

### The myosin cross-bridge cycle

Data from numerous studies of the actin–myosin interaction in solution
and in permeabilized muscle fibers have been combined to produce a
working model of the myosin cross-bridge cycle and thus the process
of chemo-mechanical transduction in muscle. Starting with a detached
myosin head to which ATP is bound (Figure 5.3), the cycle begins (Step
1) with rapid hydrolysis of ATP to ADP and $P_i$ (inorganic phosphate).
The hydrolysis products are released at an extremely slow rate, unless the
myosin binds to actin. It appears that when myosin is not bound to actin,
it can exist in one of two structural states. Either the ATP state, which is
similar to the structure with ADP bound, or in the state following hydrolysis, in which both $P_i$ and ADP are bound. This ADP.$P_i$ state is in
essence the "energized" state, in which the free energy available from
ATP hydrolysis has been stored. It is thought that the hydrolysis causes
the head of the myosin molecule (Figure 5.2) to change shape, so that the
light chain binding domain swings into a position that is approximately
10 nm away from its position in the ATP state (see Figure 5.3).
Movement back to the ADP structure (essentially the ATP structure)
occurs when $P_i$ is released. However, the release of $P_i$ is very slow unless

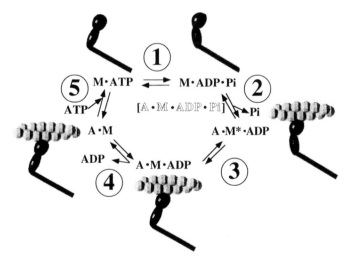

Figure 5.3. Simplified version of the actin–myosin cross-bridge cycle.
Abbreviations: M = myosin; A = actin; M·ATP = myosin with ATP bound;
M·ADP·Pi = myosin with ADP and inorganic phosphate bound;
M·ADP = myosin with ADP bound; M* = force producing ("energized") myo-
sin cross-bridge.

the myosin binds to actin. This is obviously of key importance, because the energy stored in the myosin ADP.$P_i$ state can only be recovered as force and movement if the myosin is bound to the actin filament when the $P_i$ is released.

Thus Step 2 in the force/movement producing cross-bridge cycle is the sequence of events in which myosin attaches to actin and releases $P_i$. This step involves what is commonly referred to as a transition between a weakly bound (to actin) myosin state and a strongly bound state. It is thought that the transition from the weakly bound to a strongly bound state requires the myosin structure to shift from its ADP.$P_i$ ("energized") conformation to a structure that favors the ADP conformation. Thus this is the step where the force is produced. If there is not an externally imposed load on the myosin, then the light chain binding domain will rapidly move into its ADP conformation. However, if there is a load on the myosin opposing this movement, then the rate at which the movement occurs will be greatly slowed. This movement is designated as Step 3 in Figure 5.3. The next step (Step 4) involves the release of ADP, which involves a further structural change in the shape of the myosin head. Once this occurs, ATP rebinds extremely fast, the myosin detaches from actin and the cycle is complete.

## Is myosin optimized for muscle contraction?

The distinct steps of the cross-bridge cycle seem to be rate limiting for different contractile parameters of muscle. The dynamic properties of muscle that shape its performance are the rate of force generation, the rate of relaxation, and the velocity at which muscle can shorten. The rate of force generation is limited by the rate of Step 2 of the cross-bridge cycle (weak to strong transition; Figure 5.3). The rate of relaxation in most muscles is limited not by myosin, but by the rate of removal of calcium from troponin. However, in muscles where calcium removal is extremely fast, Steps 3 and 4 of the cross-bridge cycle may have to be sped up as well. The maximum shortening velocity of a muscle is limited by how fast myosin can detach from actin, which in turn is limited by the rate of ADP release (Step 4).

It has been nearly thirty years since Barany observed that for skeletal muscles taken from a number of different vertebrates and invertebrates, there is a linear relationship between the steady-state actin–myosin ATPase activity in solution and the maximal velocity of shortening of the muscle from which the myosin is extracted. The actin–myosin ATPase activity in solution is limited by the rates of Steps 1 and 2 of the cross-bridge cycle, while maximal shortening velocity is limited by Step 4. Since these processes are limited by different steps in the cycle, parallel changes in rate constants are occurring which would tend to keep the rate of force development of a muscle (determined by Step 2) proportionate to the maximal shortening velocity, at least for muscles containing the myosin isozymes represented in Barany's study. Is this evidence for optimization of design or are the rates of Steps 2 and 4 tightly coupled as a result of the structure of the myosin motor? Recently, my laboratory has made mutations in myosin that clearly demonstrate that the two rates can be varied independently of each other. Thus Barany's correlation would seem to support an optimization of design rather than a reflection of myosin parameters that are inextricably linked.

The two most important steady-state properties of muscle are isometric energy consumption and isometric force generation. In all but the fastest muscles, the isometric energy consumption is dominated by the isometric myosin ATPase activity. In fast skeletal muscle, this is limited by Step 3, the rate at which myosin completes its power stroke under an imposed load. Remember that this rate is fast in the absence of load. This causes the isometric duty cycle (the percentage of the cross-bridge cycle for which the myosin is generating force while bound to actin) to be greater than the unloaded duty cycle and results in a greater economy of isometric force production. This phenomenon is known as the Fenn effect. As a muscle shortens at increasing velocities, the load on the cross-bridges decreases, and the rate of Step 3 increases to the point where it becomes faster than the rate of ADP release (Step 4). As noted above, it is the ADP release rate that ultimately limits the maximal shortening velocity. (In smooth muscle, and perhaps in slow skeletal and cardiac muscle, Step 4 may be slow enough to effect isometric ATPase activity, as well as limit the maximal shortening velocity.)

The amount of isometric force that a muscle produces will be a function of the percentage of the cross-bridge cycle that myosin spends in the force producing states (duty cycle), the unitary force each myosin head produces, and the number of myosin heads acting in parallel. At the level of the individual myosin molecule, it is thought that the unitary force is similar among different muscle myosin isozymes, which leaves only the duty cycle as a variable. The isometric duty cycle (and thus muscle force) will increase as Steps 3 and 4 slow and /or as Steps 1 and 2 speed up. (For fast skeletal muscle, Step 4 is likely too fast to have any limitation on the duty cycle.) Thus maintaining a constant isometric force per cross-sectional area requires parallel changes in rate constants. For example, slow skeletal muscle fibers have about the same force per cross-sectional area as fast fibers, even though their isometric ATPase activity (determined by Step 3) is slower. Thus Steps 1 and/or 2 of the cross-bridge cycle for a slow myosin must be proportionately lower than that for a fast myosin.

Indeed consistent with optimization of myosin design is the fact that the force per cross-sectional area (and thus the myosin duty cycle) is relatively constant in mammalian skeletal muscle fibers that contain a variety of different myosin isozymes (types I, IIa, IIx, and IIb). Yet it is clear that the duty cycle is not an invariant myosin parameter. Evidence that the duty cycle of the myosin motor can be changed is derived from comparisons of the myosin of fast and slow skeletal muscle with that from smooth and cardiac muscles. Smooth muscle myosin generates greater force per myosin cross-bridge than does its skeletal muscle counterpart, and it does so by increasing the duty cycle. This likely represents a better (optimized?) compromise for smooth muscle. In the case of rodent cardiac muscle, two predominate myosin isozymes are expressed. The slower of the two ($\beta$ cardiac) is common to both slow skeletal and cardiac muscle, and produces a force per cross-bridge that is similar to that produced by the fast skeletal isozymes of myosin. The faster of the cardiac isozymes ($\alpha$ cardiac) produces less force per cross-bridge than other mammalian striated muscle myosins via a shorter duty cycle. Thus the $\alpha$ cardiac myosin achieves its greater speed at the expense of force. Perhaps the $\alpha$ cardiac myosin isozyme is a better (optimized?) compromise for the rodent heart when it is working at high heart rates against normal pressures than the $\beta$ cardiac form, which is expressed when the heart is working at lower heart rates and against higher pressures.

The existence of differing duty cycles for the myosin isozymes expressed in the three mammalian muscle types is suggestive of optimization of myosin design for the tissue needs. This does not mean that the myosin motor is ever the optimal motor for any muscle. However, it would appear that within the constraints of muscle architecture and the structure of the myosin molecule, isozymes have evolved that may be optimized for specific functions. Understanding how the myosin motor is "fine tuned" to provide this optimization ultimately will require a detailed analysis of the structural domains of the myosin molecule.

102        *H. L. Sweeney*

## Further reading

Goldman, Y. E. (1987) Kinetics of the actomyosin ATPase in muscle fibers. *Ann. Rev. Physiol.*, **49**, 637–54.

Rayment, I., Holden, H. M., Whittaker, M., Yohn, C. B., Lorenz, M., Holmes, K. C. and Milligan, R. A. (1993) Structure of the actin-myosin complex and its implications for muscle contraction. *Science*, **261**, 58–65.

Rayment, I., Rypniewski, W. R., Schmidt-Bäde, K., Smith, R., Tomchick, D. R., Benning, M. M., Winkelmann, D. A., Wesenberg, G. and Holden, H. M. (1993) The three-dimensional structure of myosin S1: a molecular motor. *Science*, **261**, 50–8.

Smith, C. A. and Rayment, I. (1996) X-ray structure of the magnesium (II) · ADP · Vandate complex of the *Dictyostelium discoideum* myosin motor domain to 1.9 Å resolution. *Biochemistry*, **35**, 5404–17.

Sweeney, H. L. and Holtzbaur, E. (1996) Mutational analysis of motor proteins. *Ann. Rev. of Physiology*, **58**, 751–92.

Whitaker, M., Faust, L., Smith, J., Milligan, R. A. and Sweeney, H. L. (1995) A 35Å movement of smooth muscle myosin on ADP release. *Nature*, **378**, 748–51.

# 5.4 Matching muscle performance to changing demand

## LAWRENCE C. ROME

One of the most fascinating areas of physiology is the study of how parameters of a given system are fine-tuned to provide optimal performance under a variety of conditions. Here we examine how the mechanical properties of the muscular system are designed to power locomotion and sound production. For several reasons, the muscular system provides an exceptional model to examine the principles of physiological design. First, the performance at the level of the whole animal can be easily studied and is intuitively important. Second, because of the regular geometry of the muscle, integration from the level of the cross-bridge to whole animal movements is relatively straightforward. Third, motor performance shows a tremendous dynamic range both within a given animal (for example, locomotory muscle in toadfish operates at 1–2 Hz while its sound-producing muscle operates at 200 Hz) and between species.

In particular, maximum velocity of shortening ($V_{max}$), rate of relaxation, rate of activation, troponin affinity, SR pump density, mechanical gearing of the fibers, and myofilament lengths are known to show considerable variation. Much of the field of muscle physiology focuses on *how* this variation is achieved at the molecular level. However, in this part of the chapter, we will be principally concerned with *why* this variation occurs in the first place.

### What are the rules?

During steady activation, the force muscle generates depends on the amount of overlap between myosin thick filament and actin thin filament. It would seem sensible for the fiber gear ratio (change in body position/ change in sarcomere length) and myofilament lengths to be adjusted so that, no matter what movements the animal makes, the muscle would operate at optimal myofilament overlap, that is, where the muscle generates near-maximal force.

There is also a dynamic design consideration which takes into account that muscle shortens during locomotion. The force muscle generates is a function of $V/V_{max}$, where $V$ is the velocity of shortening. More importantly, the mechanical power that a muscle generates and the efficiency with which it generates the mechanical power are functions of $V/V_{max}$ as well. Again we might anticipate that the design parameters $V$ and fiber gear ratio would be adjusted in such a way that, no matter how fast the animal's movement, the muscle fibers would operate over a range of $V/V_{max}$ values (0.15–0.40) where the fibers generate maximal power at maximum efficiency.

### Design constraint no. 1: myofilament overlap

By using fish as an experimental model, we have found that at low swimming speeds, the red muscle (Figure 5.4A), which powers this movement, undergoes cyclical excursions in sarcomere length (SL) between 1.89 and 2.25 $\mu$m, centered around a SL of 2.07 $\mu$m. Further, we determined from electron microscopy that thick and thin filament lengths of the red (1.52 $\mu$m and 0.96 $\mu$m) and white (1.56 $\mu$m and 0.99 $\mu$m) muscle in carp are similar to those in frogs. Using the frog SL–tension relationship, the red muscle was shown to be operating over a range of SLs where no less than 96 percent maximal tension is generated (Figure 5.4A).

We then examined the most extreme movement of the carp, observed during the escape response. If the red muscle were powering this movement, it would have to shorten to a SL of 1.4 $\mu$m where only low forces can be exerted. Rather, it is the white muscle that performs this movement. Because the white muscle has a different orientation and a fourfold higher gear ratio than the red, the white muscle has to shorten to only 1.75 $\mu$m in the posterior and 1.9 $\mu$m in the front of the fish where most of the muscle is located. Thus the myofilament overlap is never far from its optimal level, even in the most extreme movements.

We have shown that the muscles of frogs also shorten over optimal myofilament overlap during jumping. In addition, it has been shown that this is also true in flying birds and galloping rabbits. It appears, therefore, that animals are designed in such a way that no matter what the movement, the muscles used generate nearly optimal forces. As such, myofilament overlap can be considered a design constraint.

## Design constraint no. 2: $V/V_{max}$

Another important constraint appears to be the ratio of muscle short-ening velocity during locomotion and maximum velocity of shortening, $V/V_{max}$. The first test of the importance of the $V/V_{max}$ was to determine whether faster fibers (high $V_{max}$) were used for faster movements so that they would operate at the same $V/V_{max}$ as slow fibers (low $V_{max}$).

We found $V_{max}$ in the carp to be 4.65 muscle lengths per second (ML s$^{-1}$) for red and 12.8 ML s$^{-1}$, which is 2.5 times higher, for white muscle (Figure 5.4, C,D). During steady swimming the red muscle is used over ranges of velocities of about 0.7–1.5 ML/s, corresponding to $V/V_{max}$ of 0.17–0.36; this is where maximum power is generated. At higher swim-ming speeds, and thus higher $V$, the fish recruited their white muscle because the mechanical power output of the red muscle declined as $V/V_{max}$ increased. It is clear from Figure 5.4C, that the red muscle cannot power the escape response, because it would have to shorten at 20 ML s$^{-1}$ (four times its $V_{max}$). But even if the white muscle were placed in the same orientation occupied by the red (that is, at the same gear ratio), it still could not power the escape response, because its $V_{max}$ is only about 13 ML s$^{-1}$. However, because of its four-fold higher gear ratio, the white muscle need shorten at only 5 ML/s to power the escape response (Figure 5.4D), which corresponds to a $V/V_{max}$ of about 0.38, which is where white muscle generates maximum power.

Thus the red and white muscle forms a two-gear system which powers very different movements. The red muscle powers slow movements, whereas the white muscle powers very fast movements, both while work-ing at the appropriate $V/V_{max}$. Given the constraint of $V/V_{max}$, to achieve a full repertoire of movement, animals must use different fiber types with different $V_{max}$ values and different gear ratios.

Additional studies have revealed that fast and slow swimming species of fish, fish swimming at low and high temperatures, and even jumping frogs also use their muscle at optimal $V/V_{max}$ (0.17–0.36).

It appears from these examples that animals use their muscles over a narrow range of myofilament overlap and over a narrow range of $V/V_{max}$ where muscle generates maximum force and maximum power with opti-mal efficiency. Therefore during evolution, three design parameters (gear ratio, $V_{max}$ and myofilament lengths) appear to have been adjusted so as to obey these design constraints no matter what movement is made. Hence, these design constraints appear to constitute two of the rules by which muscular systems have been put together.

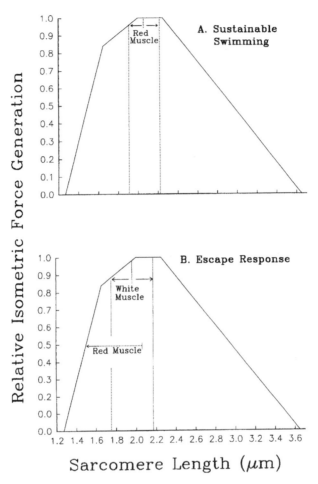

Figure 5.4. Design constraint 1: myofilament overlap. During sustained swimming (A) only the red muscle is active. The dotted lines and arrows show the SL excursion (1.91–2.23 μm). Note at the greatest excursion, force does not drop below 96 percent. If the red muscle had to power the more extreme escape response (B), it would have to shorten to 1.4 μm where it generates little tension. Instead the white muscle which has a fourfold greater gear ratio is used. In the posterior region of the fish, the white muscle shortens to only 1.75 μm (arrows and dotted line), where at least 85 percent maximal tension is generated. In the rest of the fish the excursion is smaller and the force higher. Note that because carp myofilament lengths are the same as frog, the frog SL-tension curve is used to describe the SL-tension curves for carp red and white muscles.

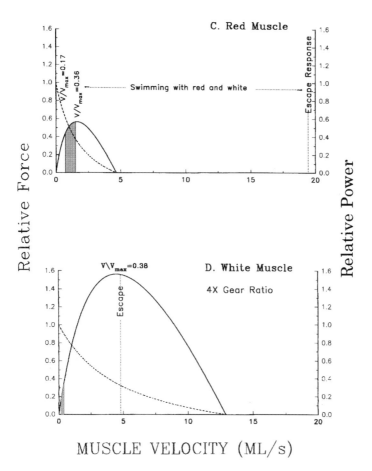

Figure 5.4.   (*continued*) Design constraint 2: $V/V_{max}$ in carp. Force–velocity (dashed) and power–velocity (solid) curves of the red muscle (C) and white muscle (D). During steady swimming the red muscle shortens at a $V$ of 0.7–1.5 muscle lengths (ML) per second (shaded region) corresponding to a $V/V_{max}$ of 0.17–0.38 where maximum power and efficiency are generated. The red fibers cannot power the escape response because they would have to shorten at 20 ML/s (C) or four times their $V_{max}$. The escape response is powered by the white muscle which, because of its fourfold higher gear ratio, needs shorten at only 5 ML s$^{-1}$, which corresponds to a $V/V_{max} = 0.38$ where maximum power is generated. (D) The white muscle would not be well suited to power sustained swimming movements (shaded), as it would have to shorten at a $V/V_{max}$ of 0.01–0.03, where power and efficiency are low.

## Design constraint for activation–relaxation kinetics

It is important to realize that the SL–tension curve and force–velocity curves are *steady-state* properties of *maximally activated* cross-bridges, and do not account for the fact that muscle must be turned on and off during locomotion. To fully understand the design of the muscular system, it is important to examine the kinetics of activation and relaxation.

Figure 5.5 shows two possible design constraints relevant to cyclical locomotion – one for maximum power output and the other for maximum efficiency. To obtain the maximum power from a muscle, the muscle must be able to instantaneously activate and relax (Figure 5.5, left panels). In this case maximum power can be generated because the muscle can be stimulated throughout shortening, and the muscle will still relax prior to relengthening. However, this requires a very fast calcium pumping rate and hence a high energetic cost.

The other alternative is to have a slow, but cheap relaxation rate (Figure 5.5, right panels). To guarantee that the muscle is mostly relaxed prior to being relengthened, the stimulation duty cycle must be shortened, and must, in fact, be shifted to start during lengthening and to end shortly after the beginning of shortening.

Thus, the question is whether evolution opted for fast-relaxing, but costly muscle or for slow-relaxing, but relatively inexpensive muscle. The only way to answer this question is to determine the characteristics of the workloops (Figure 5.5D) in the animal during locomotion.

The basic strategy we used to achieve this goal was to measure the length changes and stimulation pattern the muscle undergoes during locomotion, and then to impose these length changes and stimulation patterns on isolated muscle and measure the resulting force and power.

### *Cyclical locomotion: fish swimming*

We first let fish swim at 80 cm/s, measuring EMG and muscle length changes at four places along the length of the fish. We then removed the muscle from these regions and drove each through the length changes and stimulation pattern it sees *in vivo*, measuring the resulting force and work generated (Figure 5.6, right panels).

We found that fish muscle has been set with a relatively slow activation–relaxation rate. To ensure that the muscle is relaxed prior to being relengthened, the muscle stimulation starts during lengthening and ends

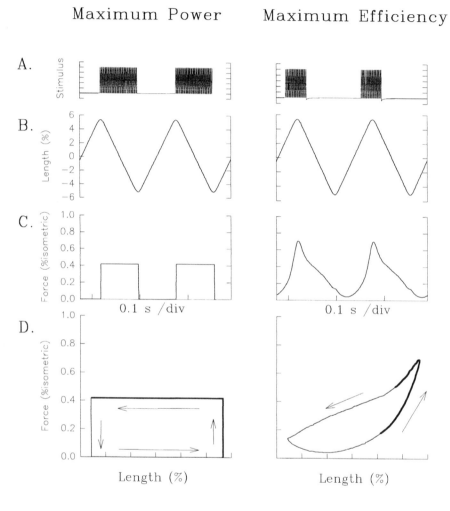

Figure 5.5.   Two possible design constraints for cyclically contracting muscle. The left panels show the stimulus (A), length change (B), force production (C) as a function of time, and workloop (force versus length; D) for a muscle that is designed with instantaneous activation and relaxation so as produce maximum power output. The right panels show equivalent graphs for a muscle designed with relatively low relaxation rates so as to produce maximum efficiency. The thickening of the workloop traces (D) represent the stimulation period.

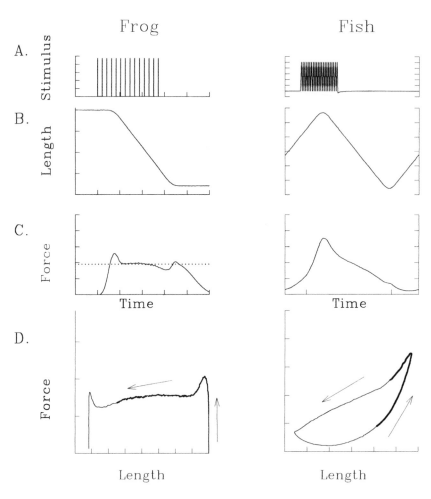

Figure 5.6.   Muscular systems designed for different functions. The left set of
panels shows the stimulus (A), length change (B), force production (C), and
workloop (D) for the frog muscle during jumping. The frog muscle is designed
to generate optimal force, and thus maximum power throughout shortening. By
contrast, the force generated by fish muscle during the shortening phase of the
swimming stroke (right set of panels) declines dramatically both due to early
cessation of the stimulus and to intrinsic shortening deactivation. This is necessary, however, if the muscle is to be relaxed by the end of shortening so that it
can be relengthened with minimal resistance. The thickening of the workloop
traces (D) represent the stimulation period. The horizontal dotted line in C of
the frog traces represents the force generated at that velocity by maximally
activated muscle.

just after the beginning of shortening (Figure 5.6, D-right panel). Thus the muscle must be relaxing through most of the power stroke.

We hypothesize that in muscles undergoing cyclical length changes, the design constraint is achieving high efficiency of power production rather than maximum power (Figure 5.5). That is, to avoid the high cost of fast $Ca^{+2}$ pumping necessary for fast relaxation, there is a compromise that reduces mechanical power production but presumably also decreases energetic cost and hence increases efficiency.

### One-shot locomotion: frog jumping

Now let us examine a very different type of locomotion, the one-shot jump of frogs. Frogs start from a crouched position with zero potential and zero kinetic energy, and in about 50 ms take off with a high potential and kinetic energy. This requires the muscles to generate a high level of mechanical power. How is this accomplished?

To generate maximum power, the frog's muscle must operate at optimal myofilament overlap, at optimal $V/V_{max}$, and must be maximally activated. We already know from the above discussion that the muscle operates at optimal myofilament overlap and optimal $V/V_{max}$.

To determine whether the muscle is maximally activated during jumping, we measured the length change and stimulation pattern of a major muscle involved in jumping, the semimembranosus, and drove isolated bundles from this muscle through the exact stimulation pattern and length changes that it undergoes *in vivo*, measuring the resulting force production (Figure 5.6, left panels). During this jumping paradigm, we found the muscle to generate the same force (Figure 5.6C) as when the muscle was shortening during the force–velocity experiments (where it is maximally activated), thus demonstrating that the muscle is maximally activated during jumping.

The comparison of the design of the frog muscular system to that of the fish demonstrates that these systems are designed differently for fundamentally different types of locomotion. Frog is designed to generate maximum power during a single shortening stroke, thus force stays high throughout shortening (Figure 5.6C, left panel). There is indeed no need for the muscle to be relaxed by the end of the shortening stroke because the muscle has sufficient time to relax before the frog hits the ground again and the muscle is relengthened. By contrast, the force the fish muscle generates during shortening declines dramatically (Figure 5.6C, right panel) due both to the early cessation of the stimulus and to intrinsic shortening deactivation. This is necessary if the muscle is to be relaxed prior to being relengthened.

## Are muscular systems optimally designed?

Clearly, muscular systems can be built to perform different functions, and to operate at different frequencies. Perhaps the most dramatic example is in toadfish: the slow twitch locomotory muscle operates at 2 Hz during swimming, whereas the swimbladder fibers, which produce sound, operate at 200 Hz. These muscles are clearly matched for their function: the swimming muscle can not mechanically produce a 200 Hz sound, and the swimbladder muscle would require 50 times more energy (due to its faster $Ca^{2+}$ and cross-bridge cycling) to produce swimming movements. But are these and other muscular systems optimally designed? This question is controversial because it comprises two narrower questions:

(1) From a purely biomechanical, physiological, and chemical kinetics viewpoint, is the system optimized for its chief function?
(2) Does natural selection optimize parameters to produce the optimal muscular system?

The first question is only concerned with whether the system is matched or not, not how it got that way (the subject of question 2), and thus can be addressed by purely physiological/biomechanical approaches.

Because of the large disadvantage associated with using the *wrong* muscle type for a given activity, it is likely that muscles are in fact tightly matched to their function. The "tightness" of this matching can be determined empirically by plotting optimal frequency of the isolated muscle power production versus the frequency at which the muscle is used *in vivo*. When sufficient data become available to make such plots, I predict that there will be little scatter around the line of unity, signifying a tight matching. Determining the evolutionary mechanisms that result in this matching, however, will require additional approaches.

## Acknowledgments

The research described here has been inspired by the insights, enthusiasm, and friendship of Professor C. R. Taylor. The work is supported by grants from NIH (AR38404) and NSF (IBN-9514383).

## Further reading

Cutts, A. (1986) Sarcomere length changes in the wing muscles during the wing beat cycle of two bird species. *J. Zool. (A)*, **209**, 183–5.

Dimery, N. J. (1985) Muscle and sarcomere lengths in the hindlimb of a rabbit (*Oryctolajus cuniculus*) during a galloping stride. *J. Zool.*, **205**, 373–83.

Lutz, G. J. and Rome, L. C. (1994) Built for jumping: the design of the frog muscular system. *Science*, **263**, 370–2.

Rome, L. C. and Lindstedt, S. L. (1997) Mechanical and metabolic design of the vertebrate muscular system. In *Handbook of Physiology*, Section 13: *Comparative Physiology*, pp. 1587–1651. Ed. W. H. Dantzler. Oxford University Press.

Rome, L. C. and Sosnicki, A. A. (1991) Myofilament overlap in red and white muscle in the swimming carp. II. Sarcomere lengths during swimming. *American Journal of Physiology: Cell Physiology*, **260**, C289–96.

Rome, L. C., Funke, R. P., Alexander, R. McN., Lutz, G., Aldridge, H., Scott, F. and Freadman, M. A. (1988) Why have different muscle fiber types? *Nature*, **355**, 824–7.

Rome, L. C., Swank, D. and Corda, D. (1993) How fish power swimming. *Science*, **261**, 340–3.

Rome, L. C., Syme, D., Hollingsworth, S., Lindstedt, S. and Baylor, S. (1996) The whistle and the rattle: the design of sound producing muscles. *PNAS*, **93**(15), 8095–100.

Sosnicki, A. A., Loesser, K. and Rome, L. C. (1991) Myofilament overlap in red and white muscle in the swimming carp. I. Myofilament lengths. *American Journal of Physiology: Cell Physiology*, **260**, C283–8.

# 5.5 Moving on land: optimizing for minimum cost

## THOMAS J. ROBERTS

Nature presents a great diversity of solutions to the problem of moving on land. Limbed vertebrates alone occur over a million-fold range of body mass, and they walk, hop, or run on two or four legs. This variation clearly affects locomotor performance; some animals are faster, stronger, or more agile than others. The metabolic energetics of running, however, is independent of animal shape or limb number, and provides evidence that terrestrial vertebrates share common solutions to the problem of economic movement.

The energy used to fuel active muscles can be determined (more or less easily, depending on the disposition of the subject) by measuring the oxygen consumption as an animal runs on a treadmill. These measurements have been performed on a great variety of runners, from mice to guinea fowl to racehorses. Two simple rules emerge: energy consumption increases linearly with speed; and small animals use more energy to move a unit body mass a unit distance (Heglund *et al.* 1982). This "cost of transport" is a regular function of body mass from mice to elephants, regardless of limb number or shape. Movement is not cheap for small animals; a mouse uses about thirty times more energy than an elephant to move a gram of its body mass a meter.

### Mechanical work of running

Many features of the mechanics of movement are also common among runners and hoppers regardless of size or shape. With each step a running or hopping animal takes, it alternately slows down in the first half of the step and then accelerates during the second half. Its body also moves downward and subsequently upward with each step. These movements involve changes in the kinetic and potential energy of the body, and

114

therefore require mechanical work. Muscle fibers could be actively lengthened to absorb mechanical energy and slow the body in the first half of the step, then actively perform work to replace this energy. Both functions require metabolic energy to fuel the working muscle.

It might be expected that the energy cost of running would be minimized when muscles perform this work most efficiently. Efficiency is the ratio of mechanical work performed to chemical energy consumed. Vertebrate skeletal muscle can operate at a maximum efficiency of 20–30 percent. If muscles perform work optimally during running, the metabolic energy consumption should be about four times the mechanical work rate in all runners. However, when the work required to lift and accelerate the body and limbs was determined by measurements of the movements of the body in various runners and hoppers, the mass-specific mechanical work rate at a given speed was independent of animal size (Heglund *et al.* 1982). A 150 kg pony and a 30 g mouse running at the same speed perform the same amount of mass-specific mechanical work, yet the mouse uses 15 times as much metabolic energy, on a mass-specific basis. Furthermore, the mechanical work rate increases curvilinearly as a function of speed, unlike the linear increase in metabolic rate. The mechanical work rate of the body does not appear to determine the metabolic energetics of running.

### Cheap running using springs

The energy cost of running is not proportional to the mechanical work rate because much of the mechanical work is done passively, by mechanisms analogous to springs and pendulums. The springs are compliant tendons that operate in series with muscles. Tendons stretched during the first half of a step release energy to help the animal take off during the second half of the step. This bounce is most obvious in a hopping kangaroo, but may be equally important in running animals. Energy can also be conserved if losses of potential energy are converted into kinetic energy and vice versa, as occurs in a pendulum. During walking this mechanism reduces the work necessary to swing the limbs and move the body.

Several lines of evidence suggest that passive mechanisms are very effective for reducing muscle work. Some of the most basic patterns of movement during running, such as the time that the foot is in contact with the ground during each step, can be predicted when animals are modeled as resonating spring-mass systems (Farley, Glasheen and

McMahon 1993). Other studies have shown that in humans, elastic energy storage in the arch of the foot alone can account for almost 20 percent of the mechanical work required in each step to lift and accelerate the body (Ker *et al.* 1987). The pendulum mechanism of energy exchange during walking is also very effective: as much as 70 percent of the energy fluctuations are recovered by exchange between potential and kinetic energy (Cavagna *et al.* 1977).

If passive mechanisms do much of the work of running, then what is the mechanical function of muscle, and how can muscle function be optimized? At first the answer might appear hopelessly complex; the intricate patterns of movement of multi-jointed limbs driven by both muscle and tendons have at least made measuring muscle function during locomotion difficult. However, the mechanical energetics of movement, as well as the properties of muscle, suggest that the optimal pattern of muscle contraction is quite simple.

During steady-speed running on level ground the average mechanical energy of the body is constant, and only a small amount of work must be done to overcome wind resistance. Almost no active mechanical work is necessary to keep the body moving forward, and passive mechanisms could, in theory, provide all of the energy to keep the limbs swinging and the body bouncing. Even perfect elastic energy storage would require active muscle force to maintain force on tendons as they stretch and recoil. Muscles use energy to produce force, even if they do not shorten to perform work. On average, muscles must produce force against the ground equivalent to one body weight (averaged over time) in order to support the body and keep the limbs from buckling.

How can the energy cost of producing this force be minimized? The answer lies in the characteristic mechanical properties of vertebrate skeletal muscle. Many of these properties are summarized by the force–velocity relation of muscle described by W. O. Fenn and A. V. Hill more than forty years ago: the force that a contracting muscle fiber develops and the energy that it uses are directly proportional to how quickly it shortens (Hill 1950). To operate at high work rates and efficiencies muscles must contract at between 20 and 40 percent of their maximum shortening velocity. These optimally working muscles, however, produce only about one-third of the force developed in an isometric contraction. This means that the optimal pattern of muscle contraction depends upon the activity: muscles should contract at intermediate shortening velocities for locomotor activities that require high work rates, but should contract isometrically and perform little or no work when force is required. The stretch and recoil of elastic tendons could allow muscles to act as economic force generators by operating isometrically.

## The cost of producing force

If muscles act primarily to produce force while passive structures do the work of running, then the cost of running should be a function of the cost of generating muscular force. Kram and Taylor (1990) found that the energy cost of hopping and quadrupedal running could be predicted from only two parameters: the amount of force produced against the ground, and the time course over which force is developed. Across both speed and size, increases in mass-specific metabolic cost were directly proportional to the amount of time available to generate force, measured as the inverse of foot contact time, $t_c$ (Figure 5.7). This time decreased with increasing running speed, and was shorter in smaller animals. The amount of force that animals must produce against the ground is, averaged over time, one body weight. Thus energy cost could be predicted from a simple relationship:

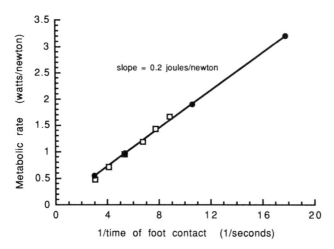

Figure 5.7.   Ponies (open squares) and mice (closed circles) use metabolic energy in direct proportion to the rate of force development, as measured from the inverse of the time of foot contact, $1/t_c$. The rate of force development accounts for differences in metabolic rate that occur with both running speed and animal size. (Reproduced from Taylor, C. R. (1994) Relating mechanics and energetics during exercise. In *Advances in Veterinary Science and Comparative Medicine* (ed. J. H. Jones), pp. 181–215. Academic Press, San Diego.)

$$\dot{E}_{\text{metab}} = W_{\text{b}} \cdot (1/t_{\text{c}}) \cdot c$$

where $\dot{E}_{\text{metab}}$ is the metabolic rate in Watts, $W_{\text{b}}$ is body weight in Newtons, $t_{\text{c}}$ is the time of foot contact in seconds, and $c$ is a constant "cost coefficient" of $0.18 \, \text{J} \, \text{N}^{-1}$.

This relationship is consistent with the energetics of force production in isolated muscle fibers. Faster rates of force development require muscles that have higher maximum shortening velocities and can develop force more quickly. Fast fibers generate the same amount of force as slow ones (when operating at the same relative shortening velocity), but their cross-bridges cycle at a higher rate. Because each cross-bridge cycle uses one molecule of ATP, fast fibers use more energy to generate force than slow fibers. Thus the increase in energy cost of running with increasing speed and decreasing size may be due to the increase in energy cost of producing force with faster muscle fibers.

### Testing the economy of force production

Direct measurements of muscle performance in wild turkeys provide more evidence that muscles act as economic force producers (Roberts *et al.* 1997). These birds are impressive athletes; they not only fly but are agile and swift runners. For a biologist interested in measuring muscle forces they display a very attractive feature – regions of tendons are calcified and as stiff as bone. This provides a site to apply miniature strain gauges, which once calibrated can measure the force output of the muscle based on how much it stretches the calcified region of the tendon. At the same time, muscle fiber length can be determined instantaneously by sonomicrometry, measuring the time it takes a pulse of ultrasound to transit between two piezoelectric crystals implanted along the muscle fibers.

We surgically implanted transducers in the lateral gastrocnemius muscle of anesthetized birds (Roberts *et al.* 1997). This muscle is an extensor of the ankle, which, unlike the human ankle, undergoes large excursions during running. Despite these large excursions, the muscle did little active shortening or work during level running (Figure 5.8). While the foot was on the ground, the muscle produced high forces but underwent very little active shortening. At the fastest running speeds the positive work done in each step was less than $3 \, \text{J} \, \text{kg}^{-1}$, or about 10 percent of what an efficiently contracting muscle can produce. Even though much of the tendon is calcified and too stiff to store significant energy, the soft compliant

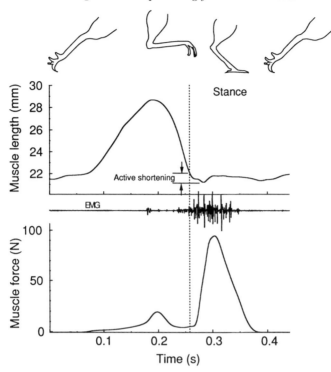

Figure 5.8.   Muscle length, activity and force in the lateral gastrocnemius for one stride of a turkey running at $3 \, \mathrm{m \, s^{-1}}$. Length changes are large in the inactive muscle during the swing phase, but small during the stance phase when the muscle is active and high forces are produced.

regions of the tendon were able to store and recover energy equivalent to almost twice the work done by active muscle fibers at all running speeds.

While we expected that tendon springs would store and recover energy, we were surprised to find that the muscle fibers themselves also acted as springs. Passive muscle resists being stretched to very long lengths; this is why it is difficult to touch your toes if you do not normally stretch your muscles. In running turkeys, force was developed passively (as indicated by the negligible EMG activity) as the ankle flexed and the muscle was stretched during swing phase. This force then helped to overcome the inertia of the foot and accelerate it towards the ground. Work that would otherwise require metabolic energy to fuel active muscle is instead performed for free when passive muscle behaves as a spring.

## Economy of running

In summary, muscular activity is economized during terrestrial locomotion with the help of biological springs. Tendons store and recover significant amounts of elastic strain energy, and can allow muscles to produce force economically. Passive muscle elasticity has rarely been considered as a useful locomotor spring, but it can potentially provide an even greater energy saving than tendon because it can function without muscle activity. The exquisite integration of these passive energy-saving structures and the active force-producing elements of muscle reduces the energy cost of producing force during terrestrial locomotion.

## Acknowledgments

The author is grateful for C. R. Taylor's advice and encouragement. Supported by NIH grant AR18140 to C. R. Taylor.

## Further reading

Cavagna, G. A., Heglund, N. C. and Taylor, C. R. (1977) Mechanical work in terrestrial locomotion: two basic mechanisms for minimizing energy expenditure. *Am. J. Physiol.*, **233**, R243–61.
Farley, C. T., Glasheen, J. and McMahon, T. A. (1993) Running springs: speed and animal size. *J. exp. Biol.*, **185**, 71–86.
Heglund, N. C., Fedak, M. A., Taylor, C. R. and Cavagna, G. A. (1982) Energetics and mechanics of terrestrial locomotion. IV. Total mechanical energy changes as a function of speed and body size in birds and mammals. *J. Exp. Biol.*, **97**, 57–66.
Hill, A. V. (1950) The dimensions of animals and their muscular dynamics. *Sci. Progr.*, **38**, 209–30.
Ker, R. F., Bennett, M. B., Bibby, S. R., Kester, R. C. and Alexander, R. McN. (1987) The spring in the arch of the human foot. *Nature*, **325**, 147–9.
Kram, R. and Taylor, C. R. (1990) Energetics of running: a new perspective. *Nature*, **346**, 265–7.
Margaria, R. (1976) *Biomechanics and Energetics of Muscular Exercise.* Clarendon Press, Oxford.
McMahon, T. A. (1984) *Muscles, Reflexes, and Locomotion.* Princeton University Press, Princeton.
Roberts, T. J., Marsh, R. L., Weyand, P. G. and Taylor, C. R. (1997) Muscular force in running turkeys: the economy of minimizing work. *Science*, **275**, 1113–15.
Taylor, C. R. (1994) Relating mechanics and energetics during exercise. In *Advances in Veterinary Science and Comparative Medicine 38A*, pp. 181–215. Ed. J. H. Jones. Academic Press, San Diego.

# 5.6 Optimization of cost of transport in swimming

## DANIEL WEIHS

The notion of optimal animal design is as old as it is controversial. This controversy results at least in part from the fact that trying to understand and explain observed behavioral patterns in terms of optimization requires several assumptions and conceptual steps – not all of which are permissible in many cases. These assumptions include the following:

(1)  that there exist one or more quantifiable dominant factors $F_d(i)$ driving the specific behavior or process;
(2)  that optimizing $F_d(i)$ results in a significant advantage to the individual or group observed;
(3)  that optimizing $F_d(i)$ does not negatively affect other parameters to a degree that results in a global reduction of fitness;
(4)  that the observed behavior or process is currently at optimal level.

Here we discuss the cost of transport, defined as the energy required to cross a unit distance, as a dominant factor in the design of the locomotor system.

### Model analysis

We select energy as the dominant factor in locomotion and separate out relatively simple activities to be analysed. Thus, we first look at long-distance directed motion (LDM) which we define as movement over distances that are at least two orders of magnitude larger than the animal's longest dimension. Another condition we place on LDM is that it is a steady, or periodic activity carried out over times much longer than that required for typical maneuvers.

The most basic case is that of an animal moving continuously over a long distance in a straight (or geodesic) line, with no energy intake. The optimization question we pose is based on the assumption that this trek is a "necessary evil," that is, it has to be done for a purpose unrelated to the area crossed, but resulting from conditions at one or either end: feeding, procreation, or evading danger.

Even this apparently simple problem has several branches. First, is the total distance traversed, $L_0$, predetermined and known to the animal? If this distance is known, one family of constraints can be set, based on energy available initially and some safety factor. This will then put limits on the cost of transport. If, however, the distance is not defined in advance, the optimization process is different as it needs to allow for maximal distance. A third possibility is when the environment en route is not uniform.

## Uniform locomotion

We start off with the least complex of the problems posed, the case of a uniform environment, and an initially undefined distance to cross. The animal's task is thus defined as minimizing the cost of transport, which translates, for a finite locomotory energy store, into maximizing the reachable distance.

We now cast the problem of minimizing the cost of transport in terms of a controllable variable, that is, the speed of locomotion $u$. The rate of working, that is, the energy required per unit time, is a function of this speed:

$$P = P(u) \equiv \frac{\partial E}{\partial t} \tag{1}$$

The energy required per unit distance, $F$, is obtained from the rate of working by recalling that the distance crossed per unit time is the speed:

$$F \equiv \frac{\partial E}{\partial l} = \frac{\partial E}{u \partial t} = \frac{P(u)}{u} \tag{2}$$

so that $F$ is also a function of $u$.

To obtain the (constant) speed $u_0$ at which $F$ is minimized, we write

$$\frac{\partial F}{\partial u} = 0 = \frac{u P'(u) - P(u)}{u^2} \tag{3}$$

where

$$P'(u) \equiv \frac{\partial P(u)}{\partial u}$$

or

$$P(u_0) = u_0 P'(u_0) \tag{4}$$

Optimization of cruising with given energy store, $E_T$, and known distance, $L$, is an example of constrained optimization. Here several factors additional to the rate of working come into play. To describe these, one needs to realize that for most cases, the rate of working $P(u)$ while moving at speed $u$ will be an increasing function; that is, $P$ grows when $u$ is increased.[1] This function does not have to be continuous, as when gait changes are observed in terrestrial animals, or when different muscular systems are employed, such as the aerobic and anaerobic systems in fish. Thus, criteria for choosing the cruising speed no longer have to be limited to the direct cost of transport, but may include time of crossing. This parameter is of importance in long-distance migrations of birds where late departure may give feeding advantages that more than compensate for the higher cost of transport resulting from a higher migration speed. Here also, safety factors in the form of reserves can come into the analysis.

### Nonuniform locomotion

A different generalization of the energy utilization as criterion for optimizing locomotion is obtained by relaxing the uniform speed requirement. The bounding flight of birds, burst and coast swimming of fish, and skipping by aquatic plankton are examples of such motions.

Returning to the basic case described by equations (1)–(4), we now attempt to find the optimal cruising speeds of specific organisms by defining the measurable quantities comprising the function $P(u)$. The hydrodynamic drag, $D$, for fish of length greater than 10 cm can be written as

$$D = \frac{1}{2}\rho V^{2/3} C_D u^2 \tag{5}$$

where $\rho$ is the density of the water surrounding the fish, $V$ is the fish volume, and $C_D$ is the drag coefficient, which is a calculable, and as we shall see later, also a controllable factor. The thrust $T$ has to be equal in

---

[1]An important exception appears in flying animals, where hovering at zero forward speed is sometimes much more costly than fast forward flight.

size and opposite in direction for the speed to stay constant so that the thrust is also proportional to the swimming velocity squared

$$T = D = \tfrac{1}{2}\rho V^{2/3} C_D u^2 \tag{6}$$

The energy per unit time required to produce this thrust is

$$P_t = \frac{1}{\eta} Tu = \frac{\rho V^{2/3} C_D}{2\eta} u^3 \tag{7}$$

where $\eta$ is the efficiency of swimming. The total rate of energy expenditure also includes the energy for maintenance (the standard metabolic rate $M$) so that

$$P = P_t + M = \frac{\rho V^{2/3} C_D}{2\eta} u^3 + M \tag{8}$$

By definition $M$, $\rho$ and $V$ are not functions of the velocity. The swimming efficiency $h$ has been measured for several species. It has the general form

$$\eta = Au^\alpha \tag{9}$$

where $A$ and $\alpha$ are constants (that is, the efficiency increases with swimming speed to the power $\alpha$).

We now limit ourselves to aerobic swimming by body oscillation. The drag coefficient varies under these conditions only with velocity, due to a dependence on the Reynolds number

$$C_D = C_{DC} Re^{-\delta} = C_{DC} \left(\frac{l}{\nu}\right)^{-\delta} u^{-\delta} \tag{10}$$

where $l$ is the fish length and $\nu$ the kinematic viscosity of water. Combining all factors independent of the velocity into a constant $C_1$, and writing

$$\beta = \alpha + \delta \tag{11}$$

we obtain

$$P(u) = C_1 u^{3-\beta} + M \tag{12}$$

so that the optimal speed, which fulfills condition (4) is obtained from

$$C_1 u_0^{3-\beta} + M = u_0(3 - \beta)C_1 u_0^{2-\beta} \tag{13}$$

or

$$M = (2 - \beta)C_1 u_0^{3-\beta} = (2 - \beta)P_{t0} \tag{14}$$

We thus see that even without having accurate data on a given species, a general estimate for the power required to move at the optimal speed defined by minimum cost of transport is

$$P_{t0} = \frac{M}{2 - \beta} \qquad (15)$$

or

$$u_0 = \left( \frac{M}{2 - \beta} \frac{2l^\delta A}{\rho \nu^\delta V^{2/3} C_{DC}} \right)^{\frac{1}{3-\beta}} \qquad (16)$$

We see that the simple relation between power for optimum speed and resting metabolic rate becomes much more complicated when actual optimal speeds are required. Standard metabolic rate is usually proportional to the mass $m^{0.75}$, and most fish species are roughly neutrally buoyant, so that $\rho_{\text{fish}} - \rho \ll \rho$. Weihs (1977) discussed the length dependences of the other factors in (16) showing that $\delta \approx 0.2 \pm 0.1$ and $\alpha = 0.85 \pm 0.15$. From (14) we thus see that $\beta \approx 1$ so that

$$0.95 P_{t0} = M \qquad (17)$$

and thus, for fish moving freely in the wild, the average rate of working will be approximately twice the standard metabolic rate.

Translating this into actual optimum swimming speeds is much more complicated as additional factors such as ambient temperature and stress affect the parameters appearing in equation (16). Using the data available for sockeye salmon (Priede 1985), the optimal velocity can be written as

$$u_0 \propto L^{0.43 \pm 0.07} \qquad (18)$$

The coefficient of proportionality in equation (18) is found from data on 30-cm-long salmon (Priede 1985), which shows that the rate of energy consumption grows with swimming speed as

$$\log Y = 1.61 \pm 0.34u \qquad (19)$$

where $Y$ is the oxygen consumption in $\text{mg kg}^{-1} \text{h}^{-1}$ and $u$ is the swimming speed in body lengths per second. From equation (19), $M = 40.74 \, \text{mg O}_2 \, \text{kg}^{-1} \, \text{h}^{-1}$ so that from equation (17) $P_{t0} = 38.70 \, \text{mg O}_2 \, \text{kg}^{-1} \, \text{h}^{-1}$. Thus, $u_0 = 27.6 \, \text{cm/s}$ in equation (16) and the coefficient in equation (18) is found as 0.46.

The variation of predicted optimal speed with size is shown in Figure 5.9, together with data collected for sharks that are an order of magnitude larger than the salmon used to calibrate the equation. The close

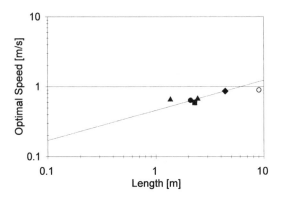

Figure 5.9.  Measured volitional swimming speeds and comparison to the calculated optimal swimming speed. The calculated optimum speed is shown by the full
line. Measured speeds for different large shark species are shown by symbols: ▲,
Mako, in the wild (Carey, pers. comm.); ◆, White, in the wild (Carey, pers.
comm.); ○, Basking, in the wild (Bone, pers. comm.); ■, Bull, in Sea World
tank (Weihs et al. 1981); ●, Sandbar, in Sea World tank (Weihs et al. 1981).

correlation between these results and the predicted curve lend credibility
to the model.

## Discussion

The models above all assume continuous variation of the parameters in
$P(u)$ with velocity. This, however, is only true over a limited range of
speeds. Animals have found ways of improving performance at speeds
very different from the aerobic cruising optimum above. This is most
easily understood in terms of gait changes in terrestrial locomotion;
applying these principles to swimming requires consideration of the
differences resulting on the one hand from the hydrodynamic drag
considered above, and on the other hand the fact that most, if not
all, weight in water is countered by hydrostatic lift. Thus several different propulsive systems have developed in swimmers, these being applied
at different life-stages and at different speeds or for different purposes
at a given time. These gait changes result in a change in the numerical
value, and even the functional form, of the drag coefficient $C_D$ in
equation (11).

Gait changes in fish can be discerned both by different propulsive
organs being applied – changing from paired fin locomotion at low
speeds, to body and caudal fin continuous oscillation, to periodic

switching from oscillatory swimming to burst and coast swimming, each of which reaches maximum efficiency in terms of costs of transport for a different speed range. Also, swimming in currents, and maneuvering will result in different fin applications, again resulting in changes in $C_D$.

A sophisticated adaptation to minimizing the cost of transport at speeds different from the simple optimum is the two-stage motion, as exemplified by burst and coast swimming and bounding flight. The basic concept here is to utilize the fact that in aerial and aquatic loco-motion one can cross significant distances without using any locomo-tory energy by using existing stores of kinetic or potential energy to overcome dissipation due to drag. This is usually not possible in hor-izontal terrestrial locomotion due to the large friction drag in contact with solids.

Kinetic energy used to counter drag results in slowing down so that this can only be a temporary respite. Similarly, using potential energy (for birds, or negatively buoyant aquatic creatures) results in height changes. Thus, after a period of thrustless motion, a period of motion with higher thrust must be used by the animal to regain the energy level (speed or height) originally occupied. An advantage in terms of total cost of trans-port can be obtained in various ways, by employing such two-phase locomotion patterns, obtaining a lower average drag coefficient over the whole cycle. This is typical of burst and coast swimming and bound-ing flight. When actively moving by body and tail oscillation, or wing flapping, the drag is much larger (by a factor of up to five) than when just gliding. Thus, if the gliding phase is long enough, the average drag over the whole period can be less. This type of strategy can be used to reduce average drag to obtain a local minimum of cost of transport, in a range of speeds far from the global optimum.

## Further reading

Lighthill, M. J. (1975) *Mathematical Biofluiddynamics*. Society for Industrial and Applied Mathematics, Philadelphia.
Priede, J. M. (1985) Metabolic scope in fishes. In *Fish Energetics – New Perspectives*, pp. 33–64. Ed. P. Tytler. Croom-Helm, London.
Webb, P. W. (1994). In *Mechanics and Physiology of Animal Swimming*, pp. 45–69. Eds. L. Maddock, Q. Bone and J. M. V. Rayner. Cambridge University Press, Cambridge.
Webb, P. W. and Weihs, D. (Eds.) (1983) *Fish Biomechanics*, pp. 339–71. Praeger, New York (see ch. 11.)

Weihs, D. (1977) Effects of size on sustained swimming speeds of aquatic
    organisms. In *Scale Effects in Animal Locomotion*, pp. 33–38. Ed. T. J.
    Pedley. Academic Press, London.
Weihs, D., Keyes, R. S. and Stalls, D. M. (1981) Voluntary swimming speeds
    of two species of large carcharhinid sharks. *Copeia*, **1**, 219–22.

# 6

# Design of cells for metabolism

## 6.1 Overview

STAN L. LINDSTEDT

When the concept of symmorphosis is extended to cellular metabolism, we shift our focus from the economical design of macroscopic to microscopic structures. In this case, one might expect to find that the relative activities of all of the cell's enzymes involved in any biochemical pathway to be "tuned" both to one another as well as the cell's absolute needs. Does natural selection favor this "optimization" of types and amounts of enzymes or are there "rate-limiting" steps suggesting that many enzymes are present in over-abundance?

One of the most significant discoveries of recent years is that metabolic pathways may operate more like bucket brigades than rivers with single "rate-limiting" bottlenecks. This concept of shared metabolic control is a view consistent with the concept of symmorphosis. That view is now refined even further to suggest that when enzymes are in their *in vivo* positions within the cell, as opposed to taken out and suspended *in vitro* in solution, the overall rate and capacity of metabolic processes is significantly altered. Overall reaction rates depend on the specific conformational positions and relationships among all the contributing enzymes. The cell is not a "bag of enzymes"; the enzymes are situated in a way that their activities vary as a function of their absolute and relative positions within the cell. The result is that pathways are controlled as units, not as single enzymes.

Likewise, the enzymes themselves are not structurally identical across species. Hundreds of different isoforms may exist for important proteins involved in, for example, ATP supply and demand pathways. These isoforms could be critical in setting the overall reaction rates to be tuned to; that is, "optimized" functionally reflecting current cellular requirements. For example, as fast and slow twitch muscles vary greatly in their ATP requirements during maximal activity, in both fiber types ATP produc-

129

tion must be perfectly balanced to ATP utilization. As these fiber types also vary in their isoform assemblages, it is concluded that these isoforms contribute to regulation.

When we consider demands that vary beyond fiber types within a given animal (for example, comparing hummingbird or even insect flight muscle with locomotor muscles from a reptile or cat), we can test the predictions of symmorphosis with a particularly robust ratio of signal to noise. Do any of these muscles have an over-abundance of enzyme, or are enzyme activities always titrated to demand?

Finally, this section of papers concludes with a comparative examination of muscle structure and energetics. Because both the metabolic and contractile structures of muscle are highly conserved, it should be possible to use a comparative approach to test if symmorphosis applies to the minimum cost of muscular contraction and if energy supply (ATP resynthesis) is matched to energy utilization (instantaneous ATP utilization) in aerobically functioning muscle. In some of the highest energy-demanding muscles the hypothesis of symmorphosis is supported in as much as the balance of energy supply and demand is similar in muscles as diverse as rattlesnake tailshaker muscle and insect flight muscle.

# 6.2 Molecular symmorphosis, metabolic regulation, and metabolons

## PAUL A. SRERE

Weibel and Taylor proposed that organisms were designed in a way such that the components of a physiological system were matched so as to economically optimize the system. They named this hypothesis symmorphosis. Diamond extended this idea to the enzyme level and referred to this as "microsymmorphosis." In the spirit of today's popular scientific nomenclature, I would like to rename that concept as "molecular symmorphosis," and in this part of the chapter I will present some of the known metabolic and enzymatic evidence that supports this idea.

### The plasticity of metabolic sequences

Cells and organisms can adapt their metabolic activity at the enzyme level to cope with changing nutrition, endocrine factors, overall changes in physical activity, or as a result of many other environmental or physiological stimuli. As examples of this behavior, the enzymes of the Krebs tricarboxylic acid (TCA) cycle and electron transport (ET) (Figure 6.1) will all increase in exercised muscle and in the muscles of animals injected with the thyroid hormone $T_3$.

Pette *et al.* (1962) observed that the Krebs cycle enzymes and the proteins of ET constitute a constant proportion group of proteins. That is to say, a comparison of the $V_{max}$ activities of these enzymes from tissue to tissue shows that their amounts (judged from their activities) remain in a rather constant ratio to each other.

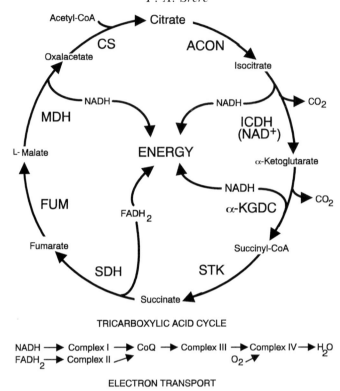

Figure 6.1. The sequence of reactions of the Krebs tricarboxylic acid cycle and the sequence of the reactions of the associated electron transport chain. The abbreviations used are CS, citrate synthase; ACON, aconitase; ICDH(NAD$^+$), isocitrate dehydrogenase (NAD-specific); $\alpha$-KGDC, $\alpha$-ketoglutarate dehydrogenase complex; STK, succinate thiokinase; SDH, succinate dehydrogenase; FUM, fumarase; MDH, malate dehydrogenase.

## Metabolic regulation

### *Classical control concepts*

Current biochemical concepts of metabolic regulation state that the activity of enzymes can be regulated by their amount, the concentration of their substrates, and the concentration of effector molecules (nonsubstrate metabolites). In a metabolic pathway composed of a linear sequence of enzymes it is believed that one (possibly two) enzyme is a rate-controlling step. Further, the rate-limiting step is believed to occur at a branch point in metabolism, that is, where a unique metabolic sequence begins.

The effectors of these rate-controlling enzymes are usually metabolites distal in the metabolic pathway, and their effects on the enzyme are in a direction that makes metabolic sense, i.e. a product of the pathway would shift the position of the velocity response curve to result in a lowering of activity.

### Distributive control concepts

Metabolic regulation according to the new school (it is now around twenty years old) was based originally on a theoretical model of a linear metabolic sequence. It is first assumed that the pathway is in steady-state such that the flux of metabolites, $\Delta J$, is constant over time. Under this condition, each enzyme catalyzed step must proceed at a rate identical to the overall flux and must be capable of regulating the rate of the sequence.

Each step can be characterized by (1) the quantitative effect, a change in its activity on the flux of the system (flux control coefficient); and (2) the sensitivity of each enzyme to its substrate (elasticity coefficient).

A number of excellent papers have determined the control coefficients for several different metabolic pathways, including glycolysis, electron transport, and fatty acid oxidation. The results of these experiments can be summarized as follows: (1) control is distributed along a pathway (that is, there is no one rate-limiting step); (2) control can vary among different steps depending on the way the flux is altered; (3) the "classically" defined control steps may under certain circumstances have little or no control; and (4) so-called equilibrium reactions may have significant control in a sequence.

### Metabolons: complexes of sequential metabolic enzymes

Another metabolic concept is that of complexes of sequential metabolic enzymes (metabolons). Although it is accepted that there exist multienzyme proteins, multiple enzyme activities on a single polypeptide chain (fatty acid synthase) and tight complexes of separate sequential activities (pyruvate dehydrogenase complex), there is little or no acceptance of the good evidence reported for *in situ* complexes of so-called soluble enzyme systems such as glycolysis, the Krebs TCA cycle, amino acid synthesis, nucleotide synthesis, photosynthesis, and many other metabolic pathways. These pathways of small molecule metabolism differ from the generally accepted metabolic complexes for the synthesis of protein, DNA, and RNA, in that they lack the complete processiveness of the pathways of macromolecular synthesis. The dissociation constants for these complexes are relatively high, and disruption of the cell and dilution of the ligands results in their dissociation.

When whole *Euglena* cells are placed in a centrifuge and subjected to high *g* forces, then the cell components separate with the heaviest particles and organelles at the bottom and a clear layer at the top, a fraction known as the cytosol. Surprisingly, an examination of this fraction by micro-spectrophotometry revealed no enzymes or proteins to be present.

When mammalian cell membranes are removed or the cells made permeable to large molecules, surprisingly little protein is lost from the aqueous cytoplasm.

It is of course naive to think of the cellular interior in terms of solution chemistry. The model should be of a thixotropic gel since the macromolecular concentrations are so high. The aqueous cytosol is about 30 percent protein while the mitochondrial matrix is about 50 percent protein; concentrations which are in the same range as that in protein crystals.

These considerations apparently give a firm basis to the discrepancies of observations in intact cells and the observations on the same cells disrupted and diluted by experimental treatment.

### The Krebs TCA cycle metabolon

We and others have studied the Krebs TCA cycle (Figure 6.1) metabolon for about twenty years. Such a complex was suggested by Green and his colleagues about thirty years ago; however, the complex they isolated was the mitochondrion. Subsequent studies on disrupted mitochondria seemed to indicate that the Krebs TCA cycle enzymes (except succinate dehydrogenase) (SDH) were soluble matrix proteins. I was led to reconsider this subject when I calculated that the bulk mitochondrial matrix concentration of oxalacetate (OAA), a substrate for citrate synthase (CS), was too low to sustain the observed rate of aerobic oxidation in that cell. I hypothesized that the OAA existed in a microenvironment between malate dehydrogenase (MDH) and CS, giving the OAA an apparently higher thermodynamic potential than its bulk matrix concentration would indicate. In the ensuing two decades, using a variety of approaches, I have studied the possibility that such a metabolic complex exists *in situ*.

### Interaction of pure enzymes

Using physical biochemical methods, it can be shown that pure sequential Krebs TCA cycle enzymes interact specifically. As an example, we have shown that CS interacts with mitochondrial MDH, mitochondrial aconitase (mACO) (citrate utilization), thiolase (acetyl-CoA supply), pyruvate dehydrogenase complex (PDC) (acetyl-CoA supply), and citrate transporter (citrate utilization). It does not interact with the cytosolic isozymes of MDH (cMDH) or cytosolic ACO.

### Interaction of Krebs TCA cycle enzymes with the inner membrane

It has been observed that the Krebs TCA cycle enzymes bind to inverted (inside out) vesicles of mitochondrial inner membranes, but not to the outer surface of the inner membrane, or to other cellular membranes.

### Kinetic effects of Krebs TCA cycle complexes

A number of kinetic effects have been observed on complexes of the Krebs cycle enzymes. When the complex of PDC and CS was examined, a lowered $K_m$ for acetyl-CoA for the CS reaction was observed. Also the rate of OAA reduction by MDH was enhanced by the addition of CS.

Other coupled reactions, besides malate to citrate, for Krebs TCA cycle enzymes were also examined. These included the oxidation of fumarate and the oxidation of isocitrate. For these studies the lightly sonicated mitochondrial preparation (metabolon) was used. This preparation contains all the enzymes of the Krebs TCA cycle in a bound form. For comparison, a thoroughly sonicated mitochondrial preparation, one in which all (except SDH) were free, was used. The results indicated that in every case the metabolon gave faster rates than the solubilized enzyme system. Such results would be consistent with a channeling mechanism in those parts of the Krebs TCA cycle.

### Orientation conserved transfer (OCT): channeling in intact cells

All of the evidence presented thus far concerning a complex of Krebs TCA cycle enzymes and channeling within the complex is derived from studies on isolated enzymes and broken mitochondria. It is necessary to see whether or not complexes existed *in situ* and if channeling occurred *in situ*. It should be realized that the lack of channeling does not indicate the absence of a complex, but channeling may be one possible function of a metabolon.

Analysis of the labeling pattern of the Krebs TCA cycle showed, even with OCT, randomization of label occurred with multiple turns of the cycle. Thus, the early long-term randomization results could not distinguish between free rotation of succinate and fumarate and OCT.

These results lend additional evidence to the *in vitro* evidence that a complex of Krebs TCA cycle enzymes exists *in vivo*. They do not supply definitive evidence concerning regulation or the metabolic importance of this metabolon. The evidence for a complex of the cycle enzymes supports the notion of symmorphosis at a molecular level since it supports the idea that this system has a balance in the stoichiometry of its components without the existence of one "rate-limiting" component.

## Regulation of Krebs TCA cycle metabolon

### *A classical approach*

Almost every enzyme of the cycle and electron transport has been named at one time or another as the regulatory step for the Krebs TCA cycle. In fact, one of the most popular regulatory steps is not an enzyme of the cycle but an auxiliary multienzyme complex (PDC) that can be regulated in many ways. It is an allosteric enzyme which can be inhibited by phosphorylation and reactivated by a phosphatase. It is thus responsive to the phosphorylation potential of the cell which can be considered the final product of the Krebs TCA cycle–ET system.

For many years the CS reaction, the first step of the cycle, was said to be the regulatory step because OAA, one of its substrates, existed in a low concentration so that this step became rate limiting. It has also been observed that CS is inhibited (competitive with acetyl-CoA) by ATP and not ADP so that regulation by phosphorylation potential was proposed. Since succinyl-CoA, a cycle intermediate, is a competitive inhibitor for the CS substrate acetyl-CoA, it has also been suggested as a regulator of CS and therefore the activity of the Krebs TCA cycle.

In addition, there have been suggestions that pyruvate carboxylase and AAT can be regulatory by supplying OAA to the cycle. One must not forget that the Krebs TCA cycle depends on the operation of ET chains so that the changes in the rate of that process, by whatever mechanism, would perforce regulate Krebs TCA cycle flux. There is evidence from many different systems, whole animals, perfused tissues, cells, and mitochondria to support most of the regulatory mechanisms cited above as being involved in the regulation of Krebs TCA cycle activity. No consensus exists among workers in the field who adhere to classical regulatory control ideas as to which component is rate limiting.

### The distributive control approach

It is apparent that the data presented above fit in with the theory of distributive control (that is, control can exist at many steps and can vary according to the conditions of the cell). This theory presents a mathematical framework and suggests experimental approaches which allow a testing of the hypothesis. This is the reason why the theory is superior to classical considerations of metabolic regulation which offer no coherent framework for testing and revision.

The greatest problem one faces in testing for distributive control is the enormous amount of very precise data one must collect in order to generate meaningful flux control coefficients. In addition, if the metabolic pathway is complex, then one must examine a few sections of the pathway at a time.

In spite of these difficulties, the Krebs TCA cycle–ET system has been examined by several laboratories regarding distributive control. The first study by Westerhoff and his colleagues examined succinate oxidation by mitochondria. They found control distributed in the dicarboxylate carrier, the adenine nucleotide translocator, and hexokinase.

The most complete study on the Krebs TCA cycle–ET system has been carried out by Veech and his co-workers using perfused rat hearts. Their results are very interesting in that they show the increase in cardiac efficiency with insulin and ketone bodies to be a result of changes in mitochondrial substrate levels and cofactor ratios. These are reflective in a reduction of $NAD^+$ and oxidation Q redox couples. Thus, it is possible at this stage to hypothesize that regulation was distributed throughout the system.

Thus, the regulatory considerations of the Krebs TCA cycle, whether from a so-called classical viewpoint or from the hypothesis of distributive control, both seem to negate the idea that a single "rate-limiting" enzyme controls this metabolon and supports the hypothesis of molecular symmorphosis.

## Genetic engineering and molecular symmorphosis

The recently acquired ability to specifically alter the enzyme content of cells seems to offer the opportunity to test many ideas concerning the metabolic role of specific enzymes in terms of their amounts and, with the use of site-directed mutagenesis, in terms of their function. From the point of view of determining which enzyme of a metabolic sequence is rate limiting, this method would seem to be ideal. After one identified the rate-limiting enzyme, one could decrease its activity in a number of ways or increase its content in the cell.

One would then compare the metabolic flux in the system to the change in the rate-limiting enzyme activity. One could also change the regulatory behavior of an enzyme from an allosteric one to a Michaelian one and again observe the changes in the flux of the system.

We have performed genetic manipulations on the CS of yeast cells in order to investigate the problem of Krebs TCA cycle enzyme complexes in yeast mitochondria. Yeast cells, unlike most animal cells, have a CS isozyme (CS2) in their peroxisomes where it is part of the glyoxylate cycle that allows yeast to grow on acetate as a sole carbon source. We have used mutant yeast cells which lack either the mitochondrial CS (CS1$^-$) or the peroxisome CS (CS2$^-$). We found the CS2$^-$ cells grow as well as the parental cell on acetate. On the other hand, CS1$^-$ cells could not grow on acetate, even when CS2 was over-expressed. This indicated that glyoxylate cycle synthesized citrate was not available to the Krebs TCA cycle. This result was interpreted in the following way: when the CS1 protein was missing, the cycle metabolon was disrupted such that the cycle activity could not provide enough energy for growth with citrate from the peroxisomes.

A word of caution is necessary concerning the use of genetic engineering of metabolic processes. It has been tacitly assumed that the metabolic changes one sees (or does not see) are attributable solely to the change in the enzyme being manipulated. It is clear, however, that the change in a single enzyme results in the change of many other enzymes.

## Summary

I have attempted to show how the concept of symmorphosis originally conceived at a physiological level is easily translated to the molecular level of cell structure and function. This concept fits in with what is already known concerning induction of metabolic pathways and with the growing awareness that metabolic regulation is a distributive function in most metabolic sequential conversions.

These ideas were illustrated by our studies on the Krebs TCA cycle metabolon. I have also indicated that further testing of these ideas can be carried out with our recent ability to genetically alter the metabolic machinery of cells.

## Further reading

Brindle, K. M. (1995) *Advances in Molecular and Cell Biology*, vol. 11. JAI Press, London.

Cornish-Bowden, A. and Cárdenas, M. L. (1990) *Control of Metabolic Processes*. NATO ASI Series. Plenum Press, New York.

Kay, J. and Weitzman, P. D. J. (1987) *Krebs' Citric Acid Cycle*. Biochemical Society Symposium No. 54. Cambridge University Press, Cambridge.

Pette, D. G., Klingenberg, G. and Bucher, T. (1962) *Biochem. Biophys. Res. Commun.*, **7**, 425–9.

Srere, P. A., Jones, M. E. and Mathews, C. K. (1990) *Structural and Organizational Aspects of Metabolic Regulation*. UCLA Symposia on Molecular and Cellular Biology New Series, vol. 133. Alan R. Liss, New York.

Welch, G. R. (1985) *Organized Multienzyme Systems*. Biotechnology and Applied Biochemistry Series. Academic Press, New York.

Welch, G. R. and Clegg, J. S. (1986) *The Organization of Cell Metabolism*. NATO ASI Series A: Life Sciences, vol. 127. Plenum Press, New York.

# 6.3  Are protein isoforms requisite for optimizing regulation of ATP turnover rates?

## PETER W. HOCHACHKA

When isozymes (defined as different protein forms catalyzing the same biochemical reaction) were discovered in the early 1960s, many biochemists were at first reluctant to accept the concept that isozymes or isoforms, the generic term, had functions. Many years and much hard work later, the functionality of most isoform systems is widely accepted. A good place to begin our analysis of whether or not isoforms are involved in the regulation of ATP turnover is to consider where isoforms come from.

### Gene duplication potentials, gene families, and protein isoforms

Gene duplication is a major cornerstone of molecular evolution. With duplicate genes coming under different selective pressures, ancestral genes diverge through phylogenetic time, forming so-called gene families. Just how large the duplication potential in phylogeny is, can be illustrated by the size of gene families and by how many genes show evidence of duplication. With regard to gene family size, one of the largest I have been able to find is that for protein kinases: up to 2000 protein kinases are predicted from current molecular studies! With regard to frequency of occurrence, this process is so common in phylogeny that isoforms are now known for most cellular proteins. Their numbers seem to be correlated with their roles in physiology.

For proteins charged with intracellular structure or metabolic functions, one to two duplications is the rule, generating two to four gene products, and at least this number of isoforms. While one or two duplications is most common for proteins involved in intracellular roles, there are instances of isoform families (such as isomyosins) arising from three to four duplications, generating eight to sixteen gene products.

For proteins involved in functions at the cell margin (sensing, signal transduction, cell–cell communication), there seem to be no duplication ceilings. The examples of over 2000 protein kinases or of hundreds of isoforms of GTP binding proteins indicate (i) that supply–demand rules may apply in the numbers of duplications that can be sustained, and (ii) that specificity of cell surface sensing, signalling, and cell–cell communications requires enormous diversification and specialization of protein isoforms charged with these functions. Any two "descendent" genes can come under different selective pressures and ultimately can be adapted for different functions. For example, the GLUT genes for six mammalian glucose transporter isoforms, themselves highly specialized in terms of tissue-specificity, regulation, and catalytic properties, are members of a larger superfamily of genes whose proteins have become adapted for the transport of a variety of other hexoses and carbon compounds.

### How many muscle cell types? How many muscle "car" models?

Since cells and tissues are specific assemblies of gene products, and each product can be expressed in a different form, it is obvious that the number of different ways in which cells could be assembled in any given tissue are enormous. For skeletal muscles of vertebrates, I was readily able to find about 150 proteins documented to occur as isoforms; during development and differentiation these could be assembled into a huge number ($2^{150}$) of combinations or cell types. Instead, only two or three kinds of striated muscle cells are generally found (fast-twitch glycolytic or FG, fast-twitch glycolytic oxidative or FOG, and slow-twitch oxidative or SO, fibers). If we consider cardiac muscle as another type of striated muscle cell, then in vertebrates the number of muscle cell types whose assembly is allowed in ontogeny is about four. In extreme cases, such as hummingbird flight muscle, only one muscle cell type is allowed, which represents the most extreme restriction on the theoretical and astronomical number of cell types which could in principle be assembled. The need for this restriction, our first clue about the general importance of isoforms, may arise from the need for specific kinds of regulation.

### Strategy for assessing the importance of protein isoforms in optimizing the regulation of ATP turnover

Muscle as a tissue is unique in the large-scale changes in work rate that it is able to sustain. Resting mammalian muscles utilize and make ATP at the same rate; the resting ATP turnover rate is about 1 mol per g per min. The maximum sustained ATP turnover rate for small muscle masses is about 100 times greater. If in a given muscle the regulation of ATP turnover were fully optimized it would sustain these large changes in ATP turnover rate with minimal or no changes in the concentrations of ATP or of phosphocreatine (the latter, PCr, "buffers" [ATP] through an equilibrium reaction catalyzed by a high-activity enzyme, creatine phosphokinase, CPK). Put more quantitatively, for any given period of muscle exercise, the efficiency of regulation of ATP turnover equals

$$\frac{\Delta[\text{ATP}] + \Delta[\text{PCr}]}{\text{total ATP turnover}}$$

both numerator and denominator being expressed in μmol per gram over the time period of the work. Since resting [ATP] is usually quite high, while concentration changes are quite small, changes in [ADP], [Pi], or [inosine monophosphate] also indicate how closely ATP demand and ATP supply pathways are regulated.

The point of emphasis is that our quantification gives us a way of comparing the efficiency of regulation in muscles composed of different isoform assemblages. *The more the regulation of ATP turnover is optimized, the lower the perturbation of the high energy phosphate metabolites for a percentage change in work rate.* If different muscle types were equivalent, then the isoforms from which they are assembled clearly could not influence this regulatory characteristic; as far as regulation efficiency goes, isoforms then could be considered as redundant. In fact, the latter alternative can be ruled out on the basis of recent studies showing that in muscles of humans and other mammals the efficiency of regulation of ATP turnover is fiber type specific: highest in heart, intermediate in soleus (mainly slow fibers), and lowest in gastrocnemius (mainly fast fibers). In heart, regulation is close to perfect, with no measurable perturbation in any of the high-energy phosphate metabolites, despite large changes in ATP turnover rate (Table 6.1).

### Why is regulation not optimized in all muscles?

The answer to this riddle seems to involve a trade-off between power output and regulatory efficiency. A fundamental functional difference between fast and slow type muscles is the much higher power output achievable in the former. Fast muscles are called fast because they can flare-up ATP demand and ATP supply pathways much more rapidly than

Table 6.1. *Summary of $^{31}P$ magnetic resonance spectroscopy estimates of metabolite concentration changes during changes in work of the human gastrocnemius, the human soleus, and the heart*

|  | ATP turnover | [PCr] | [ADP] | [ATP] |
|---|---|---|---|---|
|  | Fold change[a] in | | | |
| Gastrocnemius (fast) | 20–40 | 2–3 | 3–11 | 1 |
| Soleus (slow) | 20–40 | 1–1.5 | 1 | 1 |
| Heart | 1.5–7 | 1 | 1 | 1 |

[a]Estimates from human and dog studies (Balaban 1990; Allen *et al.* 1997; Hochachka *et al.* 1996).

can slow or cardiac muscles. The fuel and pathway preferences of metabolism that are required in consequence appear to make perfect coupling between ATP demand and ATP supply impossible during large-scale work transitions.

Probably the main distinguishing features of fast-twitch muscles involve adjustments in the catalytic capacities of creatine phosphokinase/mitochondrial pathways and of glycolysis/mitochondrial pathways. CPK activity per gram of muscle is generally highest in fast-twitch muscles, lower in slow muscles, and lowest in cardiac muscle, while mitochondrial volume densities follow the opposite pattern. This helps to account for fast muscles being able to sustain quicker flare-ups of ATP turnover without depleting ATP pools, but this is achieved at the expense of phosphocreatine, which by definition means a less efficient regulatory system.

At the same time, the carbon fuel preferentially utilized by human fast muscles is believed to be glycogen. Both the amount and the isoform of enzymes of glycolysis may be adjusted relative to slow and cardiac muscles. Because of the comparatively low affinity of glycolysis for ADP, the activation of a high catalytic capacity glycolytic pathway requires higher [ADP] – and consequently lower [phosphocreatine] – on transition to high work rates than in muscles using other (usually fat preferring) pathways of energy production. Again, by definition this means decreased efficiency – regulation sacrificed or traded-off for the power-output advantages of using glycogen and a catalytically potent glycolytic pathway. Because of the apparent impact of a potent glycolytic path on the adenylates and PCr, does this then mean that optimal regulation of ATP turnover is not possible in muscles with a strong fuel preference for glucose or glycogen?

### Achieving high regulatory efficiency in glucose preferring muscles

Clues as to the nature of metabolic changes that must be incorporated into carbohydrate preferring muscles to sustain high regulatory efficiency arise from recent studies of the heart. In post-absorptive humans, the preferred fuel for the heart is plasma-free fatty acids (FFA), as in other species including the dog. In both human and dog, changes in heart work rates are sustained with no measurable perturbation of the high-energy phosphate metabolites. By our definition, the regulation of ATP turnover rate is most highly optimized; the regulatory goal of perfect balance between ATP demand pathways and ATP supply pathways is achieved. In high-altitude-adapted humans, heart fuel preferences are shifted towards carbohydrate presumably in order to gain the advantage of improved yield of ATP per mole of oxygen utilized (advantageous in hypobaric hypoxia). Based on the above analysis of human gastrocnemius, one might anticipate less efficient regulation of ATP turnover in the heart of highland natives, but this is not observed. Even if glycolytic function seems to force the CPK equilibrium to higher [ADP] and thus lower steady state [PCr]/[ATP] ratios, the new steady state is stable; and the heart is able to sustain changes in work rate with no measurable perturbation in the high-energy phosphate metabolites. Relative to the gastrocnemius, this represents a distinct regulatory advantage (probably requisite for a continuously working tissue like the heart).

What isoforms in fast muscles would have to be adjusted in order to optimize regulation? A good model for this is hummingbird flight muscle, formed almost exclusively of FOG fibers with mitochondrial volume densities (about 30 percent) approaching those of cardiac cells in other species. From earlier studies, we know that hexokinase activity is up-regulated in this muscle. These adjustments not only influence glucose catabolism, but, as recent GLUT over-expression studies with transgenic mice indicate, also potently affect deposition rates of glycogen, which often is the main carbohydrate fuel used by working fast-twitch muscles. Although there is no detailed information on isoforms in the pathway between glucose and pyruvate, it is known that the pyruvate kinase/lactate dehydrogenase ratio is drastically increased (from values of much less than one in FG fibers to over three in hummingbird flight muscles), and that the M-type dehydrogenase is depressed in favor of the H-type isozyme. Similar (but lower magnitude) changes in activities and isoforms at this region of the glycolytic path are also thought to occur in human muscles adapted for a greater carbohydrate preference in metabolism and which display tighter coupling between energy demand and energy supply pathways.

While the above examples give tantalizing indications of how isoform assembly influences regulatory efficiency, we should emphasize that we have minimal information about effects of isoform changes in ATP demand pathways on the way metabolism is regulated. Do such changes contribute to the lower regulatory efficiency of human gastrocnemius relative to the soleus? What adaptations in regulation of ATP turnover rates, if any, are required to match the single high speed $Ca^{2+}$ release channel (the alpha isoform) in the sarcoplasmic reticulum of very fast sound-producing muscles? What changes co-adapt with the superfast myosins of jaw muscles in carnivores? To date, these kinds of questions are still waiting to be explored.

## Conclusions

The answer to the question posed in our title is clearly affirmative since the efficiency of regulation of ATP turnover is cell specific: regulatory efficiency is measurably higher in slow-twitch isoform assemblies than in fast-twitch muscles, and heart regulation is close to being fully optimized (because the balance between ATP demand fluxes and ATP supply fluxes is so precise that changes in work are not accompanied by any perturbation of high-energy phosphate metabolites).

Efficiency of regulation in fast type muscles seems to be sacrificed for high power output which can be turned on quickly. In these tissues, the catalytic capacities of anaerobic metabolic pathways/mitochondrial pathways are upregulated. In ATP supply pathways, key isoforms contributing to this loss of regulatory efficiency probably include pyruvate kinases, lactate dehydrogeanses, and enzymes plus transporters in glucose and glycogen metabolism. Nothing is known about contributions from ATP demand pathways to the lower regulatory efficiency in muscles such as the human gastrocnemius.

## Further reading

Allen, P. S., Matheson, G. O., Zhu, G., Gheorgiu, D., Dunlop, R. S., Falconer, T., Stanley, C. and Hochachka, P. W. (1997) Simultaneous 31P Magnetic Resonance Spectroscopy of the soleus and gastrocnemius in

Sherpas during graded calf muscle exercise and recovery. *Am. J. Physiol.*, **273**, R999–R1007.

Balaban, R. S. (1990) Regulation of oxidative phosphorylation in the mammalian cell. *Am. J. Physiol.*, **258**, C377–89.

Hochachka, P. W. (1994) *Muscles as Molecular and Metabolic Machines*, pp. 1–157. CRC Press, Boca Raton, FL.

Hochachka, P. W., Clark, C. M., Holden, J. E., Stanley, C., Ugurbil, K. and Menon R. S. (1996) 31P Magnetic Resonance Spectroscopy of the Sherpa heart: a PCr/ATP signature of metabolic defense against hypoxia. *Proc. Natl. Acad. Sci. USA*, **93**, 1215–20.

Hunter, T. (1994) 1001 Protein kinases redux – towards 2000. *Sem. Cell. Biol.*, **5**, 367–76.

Mueckler, M. (1994) Facilitative glucose transporters. *Europ. J. Biochem.*, **219**, 713–25.

# 6.4  Muscle energy balance in sound production and flight

### KEVIN E. CONLEY and
### STAN L. LINDSTEDT

Sound production and flight are some of the most energetically costly activities for animals. This high cost results from the high-frequency muscle contractions that are required for these activities. Many small insects such as bees beat their wings in flight at rates as high as 200 times a second (or hertz; Hz), while singing in cicadas can involve muscle contraction rates greater than 500 Hz. Since these high contraction rates are maintained for extended periods in both flying and singing animals, the muscle must have a capacity for sustaining high rates of energy supply to meet these demands. What muscle properties are responsible for balancing energy supply to the extremely high rates of energy demand of flight and sound production?

   To answer this question, we examine the structures of the muscle cell that generate and use the fuel for muscle contraction, ATP. Figure 6.2 shows the structures that produce ATP (mitochondria, MT), participate in activating the muscle contraction (sarcoplasmic reticulum, SR), and generate muscle force (actinomyosin, AM) in an electron micrograph of the tailshaker muscle of the rattlesnake. This muscle rattles at rates of 90 Hz for over an hour because it matches ATP supply by the mitochondria to the combined ATP demands of actin–myosin cross-bridge cycling (myosin ATPase) and of activation by calcium cycling in the SR. One challenge of achieving an energy balance at high ATP use rates is that the structures using and supplying ATP are all contained within the cell and must therefore share the same limited space. Thus there must be a trade-off between the space taken by the mitochondria to fuel contraction versus the space available for myofibrils to generate force or for the SR to regulate $Ca^{2+}$ flux. For the rattlesnake tailshaker muscle, these three organelles occupy nearly the entire cell volume (about 30 percent each). With this spatial constraint and the large ATP demands of high-

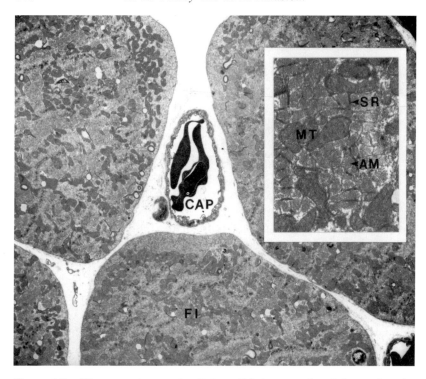

Figure 6.2. Electron micrograph of the tailshaker muscle of the rattlesnake showing the muscle fibers (FI) surrounding a capillary (CAP; original magnification, ×3000). The insert shows the three key organelles involved in the ATP balance in the cell: mitochondria (MT), sarcoplasmic reticulum (SR), and actinomyosin (AM). Insert original magnification, ×12,000.

frequency contractions, a balance of ATP supply to this demand is possible only if (1) the ATP use per contraction is minimized and/or (2) ATP supply per mitochondrial volume is maximized. In this part of the chapter, we evaluate whether such "optimization" of contractile costs and ATP supply is used to achieve muscle energy balance by studying some specific cases with high energy demand.

### The rattlesnake tail-shaker muscle

We first consider the tail-shaker muscle of the Western Diamondback Rattlesnake (*Crotalus atrox*), which produces a loud warning signal at frequencies audible to large mammals. Rattling is produced by muscle contraction at temperature-dependent frequencies of 17 Hz at 8°C body temperature up to 90 Hz at 36°C. We measured the contractile energetics of the tail-shaker muscle by non-invasive magnetic resonance spectroscopy (MR; Conley and Lindstedt 1996). The tailshaker ATPase rate per twitch is 0.02 mole ATP per gram of tissue independent of temperature or rattling frequency. Figure 6.3 shows that this ATP/twitch (solid line) is perhaps the lowest measured during flight or sound production. Costs as high as 4 μmol ATP/(g twitch$^{-1}$) are seen at the lower contraction frequencies in hovering moths (Bartholomew and Casey 1978), but these values decline to a minimal ATP/twitch in species flying at frequencies above 60 Hz. In contrast, the insert in Figure 6.3 shows that sound-producing insects in general (filled symbols) and the rattlesnake in particular (solid line) have minimized ATP/twitch values even at modest frequencies. These results support our first requirement of minimizing ATP/twitch to achieve energy balance in these muscles, but how is this minimal ATP/twitch accomplished?

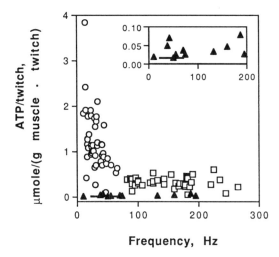

Figure 6.3.    ATP per twitch as a function of contraction frequency for the rattlesnake (bold line) and literature data: sound-producing insects (▲); and flying insects with synchronous muscles (○) and with asynchronous muscles (□). Inset is for sound-producing muscle only. Values are expressed per muscle mass for the tailshaker muscle and assume that flight and sound-producing muscles are 15 percent of body mass in insects. Frequency is the rate of tailshaking (rattlesnake), calling (sound-producers), or wing-beats. ATP utilization was determined from oxygen consumption using 22.4 ml (STDP)/μmol and a $P/O_2 = 6$. Data are from Bartholomew and Casey (1978); Casey and Ellington (1989); Conley and Lindstedt (1996); Prestwich (1994). (Adapted from Conley and Lindstedt (1996) with permission from *Nature*, **383**, 71. © 1996 Macmillan Magazines Ltd.)

### Minimizing cost of contraction

Two possibilities for reducing contractile costs are to minimize cycling of both cross-bridges (CB) and $Ca^{2+}$. Since the myosin ATPase uses one ATP per CB cycle, minimizing the number CB cycles will correspondingly reduce ATP cost. One means to keep CB cycling low is to use as few CB cycles per twitch as possible. Both insect flight and sound-producing muscles have been found to change as little as 2 percent in length during contraction, which is consistent with about one to four cross-bridge cycles per twitch (Bagshaw 1993).

A second means of reducing contractile costs is to minimize the total number of cross-bridges by reducing myofibrillar content. Both flight and sound-producing muscles have a lower myofibrillar content (30–50 percent; Casey and Ellington 1989) as compared to vertebrate locomotory muscle ( > 80 percent; Bagshaw 1993). Flight muscles apparently need a 50 percent myofibrillar content to generate sufficient lift to fly, but this is not a constraint for sound-producing muscles which often have about 30 percent myofibrils. In both cases, greatly reduced myofibrillar content minimizes the total number of cross-bridges available to cycle with each twitch. Thus, the factors contributing to a minimum ATP/twitch are (a) one to four CB cycles per twitch, and (b) reduced myofibrillar content.

The cost of muscle activation by $Ca^{2+}$ cycling is an important expense of contraction in synchronous muscles in which the nerve stimulus and muscle contraction are coupled. In contrast, asynchronous insect muscle differs from synchronous muscle in using only periodic nerve stimulation to maintain contraction. This involves a novel activation mechanism in which the stretch of each contraction cycle activates the next contraction.

By reducing $Ca^{2+}$ release to about once every ten contractions, asynchronous muscle has reduced the cost of resequestering $Ca^{2+}$ into the SR, which minimizes activation costs by largely eliminating $Ca^{2+}$ cycling. Not only does this mechanism drastically reduce contractile costs, but then the space taken by the SR can be re-allocated to other organelles, specifically to the critical ATP-supplying mitochondria. Despite the reduction in $Ca^{2+}$ cycling costs, asynchronous flight muscles have a higher ATP/twitch (mean = 0.3 μmol ATP/(g twitch$^{-1}$) than the synchronous sound-producing muscles (mean = 0.03 μmol ATP/(g twitch$^{-1}$)) for the species shown in Figure 6.3. This disparity may be due to the nearly doubled myofibrillar content of asynchronous flight muscles as compared to synchronous sound-producing muscle (50 percent versus 30 percent, respectively). In addition, there is undoubtedly a higher muscle stress and therefore greater number of CB cycles during the contraction in flying muscle to meet power requirements of lift generation. Thus the savings in $Ca^{2+}$ costs by reducing the SR may be outweighed by a higher myofibrillar content and muscle stress with the result of a larger ATP/twitch as compared to synchronous muscle.

## Maximizing ATP supply to match ATP demand

Meeting the energetic demands of contraction requires sufficient ATP generating capacity by oxidative phosphorylation in the mitochondria. Balancing ATP supply capacity to meet demand is possible by enlarging the structures that generate ATP and/or by elevating muscle temperature to increase ATP production rate. Synchronous sound-producing muscles are found to have a reduced myofibrillar content in favor of increased mitochondrial content. For example, the rattlesnake has 31 percent myofibers and equal volume proportions of mitochondria (26 percent) and SR (26 percent) for rapid calcium cycling.

In insects, such repartitioning is often insufficient to meet ATP needs of sound production ($> 200$ Hz) or may not be possible given the 50 percent myofibrillar content needed for force generation for flight. For example, mammalian mitochondria can generate about 13 µmol ATP/(ml mito $sec^{-1}$) at 30 °C (Schwerzmann *et al.* 1989). Small bees have wing-beat frequencies that exceed 200 Hz and therefore require $> 22$ µmol ATP/(g $sec^{-1}$) at a ATP/twitch of 0.11 µmol ATP/(g $twitch^{-1}$) (Casey and Ellington 1989). Clearly, even if the cell were made of 100 percent mitochondria, the ATP production rate of mammalian mitochondria at 30 °C would only meet half the ATP demand of bee flight. How then is ATP supply increased to meet the contractile costs of these high wing-beat frequencies?

Two solutions have evolved to increase the ATP supply rates in high-frequency muscle: increased structural capacity for ATP supply and/or functional rate of ATP supply. One structural adaptation is to increase the site of oxidative ATP production; that is, the inner membrane area (or cristae) of mitochondria. This membrane contains the respiratory chain enzymes concerned with oxidative phosphorylation. The area density of these cristae per unit mitochondrial volume is nearly doubled from the typical 30–40 $m^2$ $cm^{-3}$ in mammalian muscle to 60 $m^2$ $cm^{-3}$ in many flying animals such as hummingbirds and insects (Casey and Ellington

1989). This increased packing of inner membranes into the mitochondria increases the ATP synthesis capacity without changing the volume proportions of organelles within the cell. For flying bees, a 43 percent mitochondrial volume at this cristae density could generate $\sim 11.2\,\mu mol$ ATP/ (g sec$^{-1}$) and would support a wing-beat frequency of 100 Hz. However, these bees fly at twice this frequency (200 Hz), so what accounts for the additional energy supply to support this higher wing-beat rate?

A solution to this problem comes from the second method of overcoming the structural and spatial limitations of the cell: elevation of the operating temperature of the muscle to increase ATP supply rate. Mitochondria typically increase their ATP synthesis rate twofold for every 10 °C temperature increase (so called $Q_{10}$). Flying insects in general and bees in specific have muscle temperatures during flight well above ambient temperature. For example, Bartholomew and Casey (1978) report that sphinx moths (0.1–1 g body mass) range in thoracic temperature from 35 to 47 °C (average 40 °C) during hovering! Raising the body temperature of the flying bees from 30 to 40 °C doubles energy supply (with a $Q_{10} = 2$) and increases the predicted contraction frequency to that observed in nature; that is, 200 Hz. Thus the combination of increased inner membrane structure and increased temperature permit the high rates of ATP supply required to sustain high-frequency sound production and flight.

Temperature also plays an important role in asynchronous muscle because this muscle operates at a resonant frequency requiring a specific thoracic temperature for sustained muscle contraction. The cyclical contractions necessary for asynchronous flight depend on the tuned resonance of the flight apparatus (muscles and thorax) and a warm-up is required for flight or singing to elevate the thoracic temperature to a level where the muscle contracts at the resonant frequency. Thus the increase in thoracic temperature affects both sides of the energy balance: raising muscle temperature increases contractile frequency to that of the tuned system and also increases the ATP supply capacity of the mitochondria.

## Optimal muscle design?

Do these examples provide insight into whether energy balance in sound-producing and flight muscles is optimally designed? Our criteria for assessing optimality were based on the ability to achieve muscle energy balance given the high contractile frequencies and spatial constraints of the muscle cell. First, at higher frequencies, a minimum ATP/twitch is evident in a wide variety of sound-producing and flying species. Lowest in ATP/twitch among these species is the tail-shaker muscle of the rattlesnake. This minimum appears to be due to pushing the myosin to the limits of performance for CB cycles and number of active CBs per twitch. Activation costs are reduced in asynchronous muscle by circumventing $Ca^{2+}$ cycling costs by using a stretch-activation mechanism to elicit contraction. Thus our first criterion suggesting optimal design – minimizing ATP/twitch – appears to be true in these high-frequency contracting muscles.

Our second criterion for optimization – adaptation of the ATP supply capacity to match ATP demand – also appears to be true. Matching of ATP supply to demand is achieved by two means when the mitochondrial content is limited: doubling the inner membrane density of mitochondria and/or increasing the operating temperature of the muscle. These adaptations to achieve muscle energy balance satisfy our second criterion for optimal design of flight and sound-producing muscles. Thus high-frequency sound and flight muscles are designed to fully use their physiological capacity for ATP supply and demand and therefore appear to be "optimally" designed for muscle energy balance.

## Acknowledgments

We thank Larry Rome and Lee Sweeney for their comments on the concepts presented in this paper. Thanks also go to Paul Schaeffer and Marilee Sellers for the electron micrographs. Funding for this work and travel to the conference was provided by NIH AR 10853, NSF IBN 9317527 and 9306596, Northern Arizona University, and the Royalty Research Fund of the University of Washington.

## Further reading

Bagshaw, C. (1993) *Muscle Contraction.* Chapman & Hall, London.
Bartholomew, G. and Casey, T. (1978) Oxygen consumption of moths during rest, pre-flight warm-up, and flight in relation to body size and wing morphology. *J Exp Biol.,* **76**, 11–25.
Casey, T. and Ellington, C. (1989) Energetics of insect flight. In *Energy Transformations in Cells and Organisms: Proceedings of the 10th Conference of the European Society for Comparative Physiology and Biochemistry*, pp. 200–10. Georg Thieme, Stuttgart.
Conley, K. E. and Lindstedt, S. L. (1996) Minimal cost per twitch in rattlesnake tailshaker muscle. *Nature,* **383**, 71–2.

Pennycuick, C. J. (1992) *Newton Rules Biology: A Physical Approach to Biological Problems.* Oxford University Press, New York.

Prestwich, K. (1994) The energetics of acoustic signaling in anurans and insects. *Amer. Zool.*, **34**, 625–43.

Schaeffer, P., Conley, K. and Lindstedt, S. L. (1996) Structural correlates of speed and endurance in skeletal muscle: the rattlesnake tailshaker muscle. *J. Exp. Biol.*, **199**(2), 351–8.

Schwerzmann, K., Hoppeler, H., Kayar, S. R. and Weibel, E. R. (1989) Oxidative capacity of muscle and mitochondria: correlation of physiological, biochemical, and morphometric characteristics. *Proc. Natl. Acad. Sci. USA*, **86**(5), 1583–7.

# 6.5 Design of glycolytic and oxidative capacities in muscles

## RAUL K. SUAREZ

Decades of research in biochemistry and molecular biology have resulted in our current understanding of the relationships between nucleotide sequence and protein structure, enzyme structure and mechanisms of catalysis, the regulation of enzyme gene expression, and the control of flux through pathways. However, much remains unknown concerning how enzymes and mitochondria function *in vivo*. The rules that govern the design of functional capacities at the biochemical level are poorly understood.

Do metabolic pathways possess, in the words of Jared Diamond (1991), "enough or too much" enzyme? What are the relationships between enzymatic flux capacities and maximum flux rates? "Symmorphosis" predicts that capacities should match, but not exceed, maximum requirements. However, components of physiological systems are often found to be designed with built-in safety margins. Is symmorphosis an appropriate and useful hypothesis concerning the design of metabolic pathways? I have begun to address these questions by examining pathways of energy metabolism in muscles.

### Glycolysis: hypotheses versus patterns in nature

I shall first compare the enzymatic capacities for flux ($V_{max}$, equal to enzyme concentration, $[E]$, multiplied by the catalytic efficiency, $K_{cat}$) with maximum flux rates ($J_{max}$) through glycolysis. This approach is analogous to breaking down flux capacities in the respiratory system into structural elements that can be described quantitatively using morphometric units of measure and into functional elements that can be described using physiological units of measure, an approach pioneered by Weibel, Taylor and associates.

155

If God and Darwin were naive, they would have designed muscle glycolysis such that each of the $V_{max}$ values for the enzyme-catalyzed steps would equal, but not exceed, $J_{max}$. Since protein synthesis costs five ATP equivalents per peptide bond (four for peptide bond synthesis and one for active transport of each amino acid), designing muscle glycolysis to minimize [$E$] would be highly economical. However, all the enzymes catalyzing the reactions of such a pathway would necessarily have to function at $V_{max}$ when net flux is at $J_{max}$. Enzymes working under such conditions generally cannot be regulated through changes in the concentrations of allosteric effectors. At $J_{max}$, glycolytic flux would not be able to respond to changes in energy demand; rates of ATP hydrolysis (by actomyosin ATPase, $Ca^{2+}$-ATPase, and $Na^+$, $K^+$-ATPase) could not determine rates of ATP synthesis. The need to regulate flux through pathways via allosteric effectors appears to have precluded such a design. What has evolved instead is a series of enzyme-catalyzed steps, some of which are maintained far from equilibrium (mass action ratios, MAR, are far less than the corresponding equilibrium constants, $K_{eq}$), "connecting" reactions close to equilibrium (MAR values approximately equal to corresponding $K_{eq}$ values) in a manner, in Daniel Atkinson's (1977) analogy, to rivers connecting large lakes (Figure 6.4).

### Design of near-equilibrium steps

Near-equilibrium reactions, such as triosephosphate isomerase and lactate dehydrogenase (Figure 6.4), are catalyzed by enzymes occurring at $V_{max}$ values that exceed maximum rates of pathway flux, $J_{max}$, by as much as three orders of magnitude. This has often led to the impression that pathways possess excessive (that is, superfluous) capacities for flux, as well as the suggestion that no useful information can be obtained from $V_{max}$ measurements. However, metabolic biochemists have understood for decades that $V_{max}$ values need to be in large excess over $J_{max}$ to maintain near-equilibrium at such steps.

Our recent work on honeybees, *Apis mellifera*, has focused on the hexokinase (HK), phosphoglucoisomerase (PGI), and phosphofructokinase (PFK) reactions. Flying honeybees oxidize hexose sugars as their main fuels for flight. PGI activity in the flight muscles is more than 20-fold greater than the rate of glycolytic flux during flight. Figure 6.5 shows the results of a computer simulation of the PGI reaction. Using the Haldane equation, it is possible to estimate the range of $V_{max}$ values

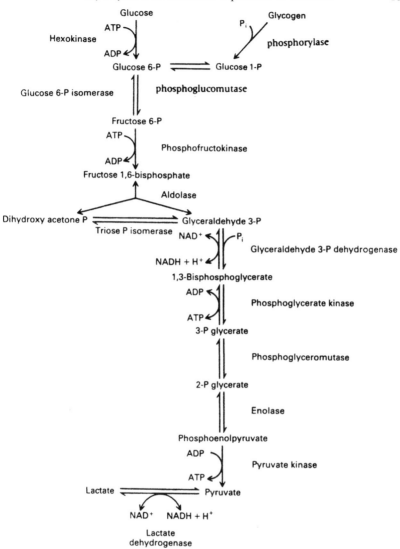

Figure 6.4. The glycolytic pathway, showing four nonequilibrium (arrows pointing in one direction) and nine near-equilibrium (arrows pointing in opposite directions) reactions. (Modified from Martin, B. R. (1987) *Metabolic Regulation: A Molecular Approach*. Blackwell Scientific, Oxford.)

required to achieve a net rate of forward flux ($J$) of 10 µmol/min, as the reaction approaches equilibrium (where $MAR/K_{eq} = 1.0$). It is apparent from this simple analysis that $V_{max}$ values greatly in excess of $J_{max}$ are actually *necessary* at reactions maintained very close to equilibrium.

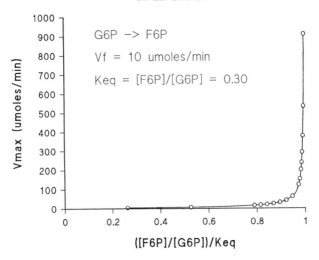

Figure 6.5.   $V_{max}$ required to achieve a net forward flux rate at the phosphoglucoisomerase step of $10\,\mu mol\,g^{-1}\,min^{-1}$ as a function of mass action ratio/$K_{eq}$, where $K_{eq} = 0.30$. The *Km* value for G6P was assumed to be 0.425 mM; [G6P] was set at 1.268 mM, and [F6P] was varied to obtain the points shown.

### Design of non-equilibrium steps

Reactions held far from equilibrium (MAR $\ll K_{eq}$) are catalyzed by enzymes whose activities *in vivo* are thought to be insufficient to bring substrates and products close to equilibrium (Figure 6.4). Unladen worker honeybees leaving their hives to forage fly with HK operating at fractional velocities (percentage of $V_{max}$) of about 75 percent, while PFK, under these conditions, works at about 44 percent of $V_{max}$. Since it is known that $\dot{V}_{O_2}$ values increase further when honeybees load up with nectar and pollen, it is likely that HK works at (or very close to) $V_{max}$, while PFK continues to operate well below saturation in bees returning to their hives after foraging. A plausible scenario is that PFK possesses the highest flux control coefficient among the glycolytic enzymes during flight. Via PGI, PFK would be able to control HK activity through mass action effects (G6P is a feedback inhibitor of insect flight muscle HK). If the fractional velocity of PFK is around 50 percent in bees leaving and returning to the hive, regulation at this step by known mechanisms (for example, changes in adenine nucleotide concentrations, F6P, F2,6-P$_2$, and so on) would still be possible at maximum rates of glycolytic flux.

Is such a close match between the $V_{max}$ values for HK and PFK and the $J_{max}$ for glycolysis observed in the muscles of other species? Figure 6.6

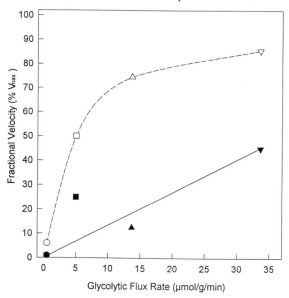

Figure 6.6.  Fractional velocities at the hexokinase (open symbols) and phospho-fructokinase steps (closed symbols) at maximal, or near-maximal, flux rates *in vivo* in rainbow trout hearts (circles), rat hearts (squares), hummingbird flight muscles (triangles), and honeybee flight muscles (inverted triangles). Original data were obtained from references cited in Suarez *et al.* (1997). These data are presented in three tables in the following web site: http://lifesci.ucsb.edu/~suarez/pnas1997tables.html.

plots the fractional velocities at which these enzymes operate versus flux rates *through these steps* under conditions of maximal (or near-maximal) rates of glycolysis in trout and rat hearts, as well as hummingbird and honeybee flight muscles. The pattern shown for HK is particularly striking, showing fractional velocities that range from very low values ($< 10$ percent) in fish hearts, to very high values that approach $V_{max}$ in hummingbird and honeybee flight muscles. PFK fractional velocities range from very low values in fish hearts, to about $\frac{1}{2} V_{max}$ in honeybee flight muscles. If honeybees possess what might be argued to be enough, and not too much HK and PFK, why should the muscles of other species possess $V_{max}$ values in such great excess over $J_{max}$? One possible explanation is that insect flight muscles function more like "complete-on–complete-off" machines than other muscle types. During the rest to flight transition in insects, flux rates go from extremely low to extremely high values. In contrast, work rates in other muscle types can be modulated incrementally over a broad range. It seems difficult to accept the idea that there might be "too much" enzyme at nonequilibrium steps in muscle glycolysis, when "optimal" $V_{max}$ values for enzymes such as triosephosphate isomerase in other systems are as much as $10^3$-fold greater than $J_{max}$. It is more reasonable to hypothesize that the apparent excess capacities at the low flux end of the spectrum are mechanistically necessary, and not simply the result of sloppy or extravagant design.

## Oxidative capacities: are muscle mitochondria created equal?

If God and Darwin were naive, mitochondria would all have been created to be the "same" in all animals, and evolution would simply have varied mitochondrial abundance in the various cell types found in different species. Economy of design, in its most simple form, would have dictated a more or less exact match between capacities for oxidative ATP synthesis and maximum capacities for ATP hydrolysis.

Instead, we find even within individual organisms that mitochondria vary in various cell types in ultrastructure and respiratory capacities. The work of Schwerzmann and colleagues (1989) on liver and muscle mitochondria shows excellent evidence of tissue-specific differences in cristae surface densities (surface areas per unit mitochondrial volume) and enzyme content per unit cristae surface area. These differences between liver and muscle mitochondria reflect differences in their metabolic roles and in maximal rates of ATP turnover.

Studies of the morphometry of muscle mitochondria in mammals have shown that these are very similar despite large differences between species in body masses and in mass-specific $\dot{V}_{O_2 max}(\dot{V}_{O_2 max}/M_b)$ values. Cristae surface densities are between 20 and $40\,m^{-2}cm^{-3}$ and during exercise at $\dot{V}_{O_2 max}$, rates of respiration per unit mitochondrial volume *in vivo* fall within a remarkably narrow range of $3\text{--}5\,ml\,O_2\,cm^{-3}\,min^{-1}$. The first hypothesis, that mitochondria are the same in all tissues, is clearly not supported by a large body of published information. However, the suggestion that muscle mitochondria are all the same, and that animals simply have more or less mitochondria in their muscles, is supported by the data collected from mammals. When animals capable of lower or higher $\dot{V}_{O_2 max}/M_b$ values are compared with mammals, it becomes apparent that what is true of mammals is not necessarily true of other taxa (Table 6.2). In particular, cristae surface densities in animals capable of higher $\dot{V}_{O_2 max}/M_b$ values can be up to double the values found in mammals. Rates of mitochondrial respiration (per unit volume) *in vivo* during exercise are higher in hummingbirds compared to mammals, and

Table 6.2. *Mitochondrial morphometry and respiration rates in locomotory muscles*

| | Mitochondria | | Respiration rate | |
| --- | --- | --- | --- | --- |
| | Volume density[a] (%) | Cristae surface density[b] ($m^2 cm^{-3}$) | $\dot{V}_{O_2}/V(mt)$ ($ml\,O_2\,min^{-1}\,cm^{-3}$) | $\dot{V}_{O_2}/S(im)$ ($\mu l\,O_2\,min^{-1}\,m^{-2}$) |
| Cuban iguana | 3 | 25 | 1 | 40 |
| Cat | 4–6 | 35 | 3–5 | 86–143 |
| Hummingbird | 35 | 58 | 7–10 | 121–172 |
| Euglossine bee | 43 | 50 | 16 | 320 |
| Blowfly | 40 | 53 | 23 | 434 |

[a]Volume densities represent the percentage of fiber volume occupied by mitochondria.
[b]Cristae surface densities represent surface areas per unit mitochondrial volume.
*Source*: Original data found in references cited in Suarez (1992).

even higher in flying insects. These higher rates cannot be accounted for simply by higher cristae surface densities, because $\dot{V}_{O_2\,max}$ values per unit cristae surface area differ by a factor of ten when comparing iguanas and flies (Table 6.2). Higher respiration rates per unit cristae surface area during exercise at $\dot{V}_{O_2\,max}$ are partly based upon increased capacities for $O_2$ delivery, but could, in addition, be due to increased enzymatic capacities for electron flux, proton pumping, and ATP synthesis per unit cristae surface area.

Our recent work on honeybees suggests that cytochrome oxidase functions closer to $V_{max}$ during flight than the homologous enzyme in mammals during exercise at $\dot{V}_{O_2\,max}$. An emerging picture is that across the Animal Kingdom, mitochondrial volume densities, cristae surface densities, cristae enzyme packing densities, and fractional velocities *in vivo* all appear to be highly variable parameters in the design of muscle oxidative capacities. Within mammals, evolution has made use of the "add-and-subtract" strategy, where mitochondria of similar morphometric characteristics scale against both body mass and $\dot{V}_{O_2\,max}/M_b$ such that maximal respiration rates fall between 3 and $5\,ml\,O_2\,cm^{-3}\,min^{-1}$. Cuban iguanas possess appropriately matched lower capacities for both mitochondrial $O_2$ consumption and $O_2$ transport. Smaller, more highly metabolically active vertebrate homeotherms such as hummingbirds have increased both mitochondrial volume densities and cristae surface densities, and are analogous to turbocharged, fuel-injected racing automobiles in their higher capacities for $O_2$ delivery and higher capacities for both fuel delivery and catabolism. Further upregulation of capacities for $O_2$ delivery is seen in insects, which lack closed circulatory systems and instead make use of air-filled trachea. Interestingly, mitochondrial volume densities and cristae surface densities in insect flight muscles do not differ much from those of hummingbird flight muscles. This suggests that upper structural limits to the upregulation of oxidative capacities may have been reached, and that the insects may have evolved a novel biochemical solution to the problem of further upregulation of mitochondrial oxidative capacities.

### Is symmorphosis a useful concept?

Symmorphosis has been described by various critics as either a truism, or as a naive idea, unsubstantiated by actual data, and unlikely to be the outcome of natural selection. Diamond (1992) has suggested that symmorphosis is a useful null hypothesis (I shall ignore the issue of whether it should be considered "null" or not). Consistent with this suggestion, Parker and Maynard Smith (1990) state that practitioners of the optimality approach in behavioral ecology do not necessarily believe in optimality, nor do they necessarily wish to prove that animals are optimal. Instead, as we have seen in this analysis, it is by formulating, testing, and refining optimality models that new questions, valuable insights, and greater understanding of mechanisms, costs, constraints, and the rules that govern evolutionary design come to light.

### Acknowledgments

I thank P. Hochachka, C. Moyes, T. West, G. Brown, L. Gass, J. Lighton, J. Harrison, B. Joos, and O. Mathieu-Costello for numerous discussions and for collaborating on various aspects of the research described in this article. My work has been funded by the Natural Sciences and Engineering Research Council of Canada and the U.S. National Science Foundation.

### Further reading

Atkinson, D. E. (1977) *Cellular Energy Metabolism and Its Regulation.* Academic Press, New York.

Diamond, J. M. (1991) Evolutionary design of intestinal nutrient absorption: enough but not too much. *News Physiol. Sci.*, **6**, 92–6.

Diamond, J. M. (1992) The red flag of optimality. *Nature*, **355**, 204–6.

Parker, G. A. and Maynard Smith, J. (1990) Optimality theory in evolutionary biology. *Nature*, **348**, 27–33.

Schwerzmann, K., Hoppeler, H., Kayar, S. and Weibel, E. R. (1989) Oxidative capacity of muscle mitochondria: correlation of physiological, biochemical, and morphometric characteristics. *Proc. Natl Acad. Sci. USA*, **86**, 1583-7.

Staples, J. F. and Suarez, R. K. (1997) Honeybee muscle phosphoglucose isomerase: matching enzyme capacities to flux requirements at a near-equilibrium reaction. *J. Exp. Biol.*, **200**, 1247–54.

Suarez, R. K. (1992) Hummingbird flight: sustaining the highest mass-specific metabolic rates among vertebrates. *Experientia*, **48**, 565–70.

Suarez, R. K. (1996) Upper limits to mass-specific metabolic rates. *Ann. Rev. Physiol.*, **58**, 583–605.

Suarez, R. K., Lighton, J. R. B., Moyes, C. D., Brown, G. S., Gass, C. L. and Hochachka, P. W. (1990) Fuel selection in rufous hummingbirds: ecological implications of metabolic biochemistry. *Proc. Natl Acad. Sci. USA*, **87**, 9207–10.

Suarez, R. K., Lighton, J. R. B., Brown, G. S. and Mathieu-Costello, O. (1991) Mitochondrial respiration in hummingbird flight muscles. *Proc. Natl Acad. Sci. USA*, **88**, 4870–3.

Suarez, R. K., Staples, J. F., Lighton, J. R. B. and West, T. G. (1997) Relationships between enzymatic flux capacities and metabolic flux rates: nonequilibrium reactions in muscle glycolysis. *Proc. Natl Acad. Sci. USA*, **94**, 7065–9.

# 7

# Lungs and gills for gas exchange

## 7.1 Overview

PIERRE DEJOURS

This chapter concerns the respiratory system, a field of physiology that played a major role in the development of the concept of symmorphosis.

The respiratory system ensures the exchanges between the ambient medium and the cells of the tissues. The respiratory media are water and air. In almost all respects, their physical and chemical characteristics differ. Whereas the $CO_2$ capacitances of both media are similar, the $O_2$ capacitance and concentration are 20 to 40 times higher in air than in water. Consequently the air-breathers breathe much less than the water-breathers to obtain the same amount of oxygen from the environment, whence $P_{CO_2}$ in the body fluids is much higher in air-breathers than in water-breathers. But the differences in the physicochemical properties of water and air – gravity, diffusivity, viscosity, heat capacitance, and so on – are enormous and, consequently, entail very marked differences in the physiology and the anatomy of the aquatic animals on one hand, and of the terrestrial animals on the other.

Oxygen and carbon dioxide exchanges take place at the interfaces between the body and the ambient medium, that is the teguments, the gills, the tracheae, and/or the lung. Oxygen diffuses from the ambient medium at the contact of the respiratory surface area to the body fluids – blood, hemolymph, or coelomic fluid – which often contain an oxygen carrier. The body fluids reach the tissue by mixing, open or closed circulation. For carbon dioxide the same phenomenon takes place in the reverse direction, that is from the cells to the environment.

Thus the general laws are valid to all animals, but there are many differences between taxa, due to their biochemical, anatomical, and physiological characteristics. One element of zoological classification is the nature of the medium in which the animals live: thus one speaks of water-breathers, of air-breathers, and of dual-breathers (amphibious animals).

Another element of classification comes from the organs (gills, lungs, and so on) situated at the interface between the animal body and the medium which characterize the modality of the external gas exchanges. For example, a tadpole is a bimodal water-breather (skin and gills); but the salamander *Necturus* breathes air through lungs, breathes water through gills, exchanges $O_2$ and $CO_2$ via the skin; thus this animal is a trimodal dual-breather. Most birds and mammals, including man, are unimodal air-breathers (no gills; negligible cutaneous respiration).

Obviously respiration is a large chapter of physiology and provides a lot of information regarding the quantitative relations between structures and functions, and thus is very important in the development of the concept of symmorphosis. Its contribution to this new field of biology cannot be developed as it should be. By necessity, one is obliged to take examples to illustrate this contribution.

C. W. Hsia studies the adaptation of lung to capacity of gas transport. How does the remaining lung, after partial pneumonectomy, adapt to fill its function of gas exchanger?

J. Maina compares the relation between function and structure in two groups of aerial vertebrates, namely by the parabronchial lung in birds and the alveolar lung in bats. Why is the mass of the largest flying bat limited to 1 kg?

E. Taylor draws attention to the very important fact that the gills of aquatic animals are, beyond gas exchangers, the seat of ionic exchanges and of ammonia excretion. How are all these functions performed altogether?

D. Randall considers the special problem of water-breathers in excreting $CO_2$, a substance which is not excreted as such, but which involves carbonic acid and the conjugate bases $HCO_3^-$ and $CO_3^{-2}$. The concentration of carbonic anhydrase is very high in the gills. The relation between $O_2$ and $CO_2$ exchanges is very important to consider since in the trout, a teleost, large Root and Haldane effects are present, whereas in a elasmobranch like the dogfish there are no Root effects and only a low Haldane effect.

## Further reading

Dejours, P. (1988) *Respiration in Water and Air. Adaptations – Regulation – Evolution*. Elsevier, Amsterdam.
Taylor C. R., Karas, R. H., Weibel, E. R. and Hoppeler, H. (1987) Adaptive variation in the mammalian respiratory system in relation to energetic demand. *Respir. Physiol.*, **69**, 1–127.

Weibel, E. R. (1984) *The Pathway for Oxygen: Structure and Function in the Mammalian Respiratory System.* Harvard University Press, Cambridge, MA.

Weibel E. R. and Taylor, C. R. (1981) Design of the mammalian respiratory system. *Respir. Physiol.*, **44**, 1–164.

## 7.2 Limits of adaptation in pulmonary gas exchange

CONNIE C.W. HSIA

A diffusion limitation of $O_2$ uptake in the lung can develop during exercise under the following conditions: (a) a low driving pressure (that is, alveolar hypoxia); (b) altered alveolar–capillary membrane (that is, high tissue resistance as in pulmonary fibrosis and edema, or reduced surface as after major lung resection); (c) a high maximal cardiac output as induced by exercise training (elite athletes) or genetic endowment (dogs and horses); and (d) inhomogeneous distribution of blood flow within the capillary bed. It is believed that in the average human or in most sedentary mammalian species, diffusive pulmonary gas exchange does not normally constitute a major limiting factor to exercise performance at sea level because the estimated capacity for ventilation and gas exchange is much greater than that for cardiovascular $O_2$ delivery. This observation, however, reflects the deconditioned state of the average human subject. Cardiovascular capacity for $O_2$ delivery can be improved through fitness training so that in elite athletes cardiovascular capacity is well matched to the capacity for gas exchange, and a diffusion limitation can become evident.

There are essentially three mechanisms by which gas exchange can be maximized in normal and disease states: (a) by hyperventilation to increase alveolar $P_{O_2}$; (b) by optimizing the position of the oxyhemoglobin dissociation curve; or (c) by reducing the alveolar–capillary resistance to diffusion. The latter mechanism can be achieved by recruitment of existing alveolar–capillary surface or volume; by reducing the molecular diffusion distance across the tissue-plasma barrier (that is, thinning of the membrane); by adding new alveolar–capillary units (that is, lung growth); or by a more homogeneous distribution of blood flow with respect to diffusing capacity. Here I shall discuss only those mechanisms that are related to the structural design of the pulmonary gas exchanger.

## Alveolar-capillary resistance to diffusion

Overall resistance of the lung to gas diffusion, which is the reciprocal of the diffusing capacity $D_L$, is composed of resistances of the tissue and the red cells arranged in series:

$$\frac{1}{D_L} = \frac{1}{D_M} + \frac{1}{\theta \cdot V_c}$$

where $D_M$ is diffusing capacity of the alveolar–capillary membrane, $V_c$ is capillary blood volume, and $\theta$ is the rate of gas uptake by red cells per milliliter of blood. The upper limit of $D_L$ is basically set by structural design parameters; the alveolar surface area and the thickness of the tissue barrier which separates air and blood determine $D_M$, and size of the capillary bed determines $V_c$. One approach to estimate $D_L$ is therefore to obtain morphometric data on these design parameters from post-mortem examination of a fixed lung. The physiologic approach to estimate $D_L$ is to measure the uptake of a tracer gas such as carbon monoxide (CO); this approach allows an estimate of variations in the parameters of $D_L$ as they occur *in vivo*. For example, as pulmonary blood flow increases, functional capillary blood volume also increases. Consequently one finds a linear relationship between physiological $D_L$ and pulmonary blood flow. One expects that at some point all available surface or volume becomes utilized in gas exchange and an upper limit of $D_L$ should be reached which may equal morphometric $D_L$. However, in normal humans or dogs, this linear relationship for CO uptake continues right up to maximal $O_2$ uptake without reaching an upper limit.

One way to further explore the limits of this relationship is to perturb the interaction between dynamic and structural basis of gas exchange by surgically removing a known fraction of lung tissue (that is, pneumonectomy). This model produces several physiologic and anatomic effects:

(a) The reduction in functioning lung units is predictable; the right lung normally constitutes 55 percent and the left lung 45 percent of total volume and receives a similar fraction of total ventilation and perfusion.

(b) After pneumonectomy, cardiac output is directed entirely through one lung instead of two, so pulmonary blood flow per unit of lung is doubled at any work load.

(c) Volume of the remaining lung expands to almost equal that of two normal lungs.

(d) If resection is sufficiently extensive, compensatory growth of the
residual lung may be triggered to restore the size of the gas exchanger.

We first performed left pneumonectomy (45 percent resection) on adult
foxhounds trained to exercise on a treadmill. When these dogs were
studied one year later, we found that $D_L$ for carbon monoxide ($D_{L_{CO}}$)
at a given level of exercise was lower than that before pneumonectomy by
only 25 percent, much less than expected from the amount of lung
resected (Figure 7.1, left upper panel). The slope of the relationship
between $D_{L_{CO}}$ and pulmonary blood flow was reduced after pneumonectomy; that is, augmentation of $D_{L_{CO}}$ by perfusion had become more
difficult. When compared with respect to the right lung alone, $D_{L_{CO}}$
after pneumonectomy was in fact higher than that for a single normal
lung, and continued to increase up to a maximal blood flow nearly twice
that in a normal right lung (left lower panel). No upper limit of $D_{L_{CO}}$ was
reached before or after pneumonectomy. This study shows that normal
capillary reserves are much greater than could be recruited physiologically. After pneumonectomy, recruitment of the remaining alveolar–
capillary bed extended beyond the normal range, up to a blood flow
nearly twice that achieved in a normal lung. Figure 7.1 also shows the
results of morphometric estimation of $D_{L_{CO}}$ in these lungs. We found
that structural diffusing capacity of the remaining right lung was
increased by about 30 percent above the control right lung (right lower
panel). Accordingly, morphometric estimates of $D_{L_{CO}}$ were similar to the
maximal physiological values obtained at peak exercise, both in the normal dog lung and that after left pneumonectomy. The increase in $D_{L_{CO}}$
after pneumonectomy was achieved by dilatation of pre-existing air
spaces which caused the alveolar surface to be stretched and the barrier
to be thinned. There was no true growth of lung tissue (Figure 7.2) except
for some enlargement of the capillary bed which caused $V_c$ to be larger.

Can an upper limit of $D_{L_{CO}}$ be reached after more extensive lung
resection? To answer this question we performed a separate study where
55 percent of lung was removed by right pneumonectomy in the adult
foxhound. Early after right pneumonectomy we found a more pronounced exercise impairment than after left pneumonectomy; maximal
$O_2$ uptake was reduced by more than 50 percent and this reduction was
associated with both ventilation-perfusion mismatch and diffusion disequilibrium. Physiological $D_{L_{CO}}$ was significantly reduced and recruitment of $D_{L_{CO}}$ during exercise was limited to a greater extent than after

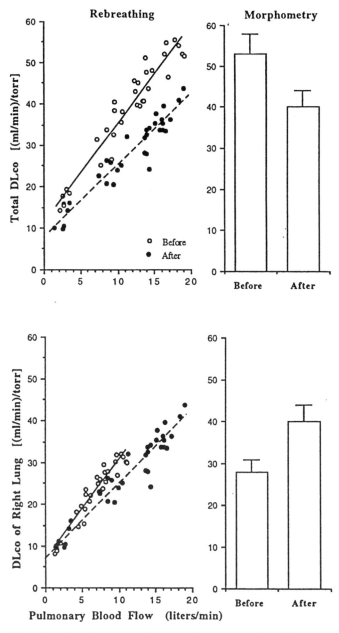

Figure 7.1.    Relationship between $D_{LCO}$ and pulmonary blood flow in adult dogs before and after left pneumonectomy measured physiologically by a rebreathing technique during exercise (left panels) and by a morphometric method (right panels). Comparsions are made with respect to both lungs (top panels) and to the right lung alone (lower panels).

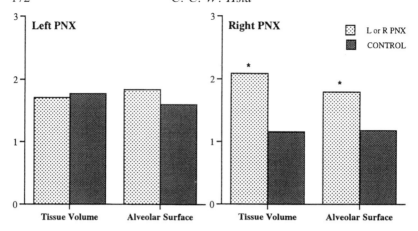

Figure 7.2. Comparison of morphometric estimates of septal tissue volume (ml kg$^{-1}$) and alveolar surface area (m$^2$ kg$^{-1}$) in the remaining lung of dogs after left pneumonectomy (PNX) (45 percent resection, left panel) and after right pneumonectomy (55 percent resection, right panel) with that in the corresponding lungs of their respective controls. Asterisks indicate a significant difference ($p < 0.05$) with respect to control animals.

left pneumonectomy. However, with time both $D_{LCO}$ and arterial O$_2$ saturation returned towards normal, although the ultimate functional compensation remained incomplete. One year after surgery, the relative magnitude of compensation in $D_{LCO}$ was greater after right than after left pneumonectomy even though a larger fraction of the lung had been removed by right pneumonectomy. Morphometric analysis demonstrates, in this instance, an increase in septal tissue volume and alveolar–capillary surface area without significant change in septal thickness in animals after right pneumonectomy compared with the same lung in control animals (Figure 7.2). Thus the more severe derangement of gas exchange after right pneumonectomy elicited compensatory alveolar tissue growth that was not elicited after left pneumonectomy. Again, physiological estimates of $D_{LCO}$ at peak exercise agree well with morphometric estimates.

### Explaining gas exchange disequilibrium

Despite the large increase in physiological $D_{LCO}$ from rest to exercise, a diffusion disequilibrium for oxygen becomes evident at heavy workloads shown by a declining arterial O$_2$ saturation and an elevated alveolar–arterial $P_{O_2}$ difference. Diffusion disequilibrium is seen even in normal dogs but is exaggerated after pneumonectomy. How can this be explained? In accordance with Fick's law of diffusion, the Bohr integral, and the shape of the oxyhemoglobin dissociation curve, end-capillary $O_2$ saturation of blood leaving the lung is a function of the ratio of diffusing capacity to pulmonary blood flow $(D_L/\dot{Q}_c)$. Because pulmonary blood flow increases at a faster rate than $D_L$, the ratio of $D_L/\dot{Q}_c$ progressively declines from rest to exercise (Figure 7.3). When this ratio drops below a critical level, the rate of diffusion cannot keep pace with the rate of perfusion and erythrocytes exit the pulmonary circulation without being fully oxygenated. This critical ratio is not usually reached at sea level in normal subjects of average fitness, because exercise is curtailed first by reaching the maximal cardiac output. This critical ratio may be reached at heavy exercise in highly aerobic subjects, and at moderate exercise when the size of gas exchange units is reduced. After pneumonectomy which significantly reduces morphometric diffusing capacity, a critical ratio of $D_L/\dot{Q}_c$ is rapidly approached during exercise and arterial $O_2$ saturation declines relatively early. Thus diffusion disequilibrium can become an important factor in determining $O_2$ loading onto blood in relation to the demand imposed by exercising muscle.

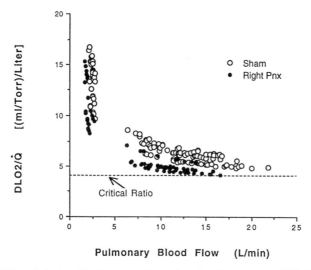

Figure 7.3. Relationship between the ratio of pulmonary $O_2$ diffusing capacity to blood flow $(D_{L_{O_2}}/\dot{Q}_c)$ and blood flow in dogs after right pneumonectomy (PNX) and in control dogs (SHAM). After PNX a critical ratio is reached below which an arterial $O_2$ saturation of 90 percent could not be maintained; that is, diffusion disequilibrium begins to limit exercise.

## Diffusion–perfusion interactions

In order to accomplish optimal gas exchange, the lung must balance several complex and dynamic processes that impose opposing demands on the system: alveolar ventilation ($\dot{V}_A$), capillary blood flow, and diffusive gas transfer. These factors must be matched for each of the many gas exchange units that make up the lung. The distribution of ventilation with respect to perfusion ($\dot{V}_A/\dot{Q}_c$) determines the alveolar $P_{O_2}$ at a given level of $O_2$ uptake. The distribution of oxygen diffusing capacity with respect to perfusion ($D_L/\dot{Q}_c$) determines the end-capillary $O_2$ saturation at a given alveolar $P_{O_2}$ and mixed venous $O_2$ saturation. To optimize capillary blood flow, the capillary bed must accommodate a high volume while maintaining a low flow resistance. To optimize diffusive gas transfer, the capillary bed must maintain a large surface of erythrocyte–tissue contact and an adequate erythrocyte transit time.

Various factors influence capillary perfusion, such as pulmonary arterial pressure, transpulmonary pressure, blood viscosity, erythrocyte shape and deformability, and the number and deformability of neutrophils which, because of their large size, may temporarily obstruct capillary segments and redirect erythrocyte traffic. The mesh-like design of the pulmonary capillary network maximizes the efficiency of perfusion by providing a nearly infinite number of possible flow pathways; progressive utilization of these pathways as flow increases maintains a low flow resistance. In doing so the system necessarily creates an extremely large capillary surface area which is beneficial for diffusion.

The disadvantage of this intricate design lies in the difficulty of providing a uniform distribution of blood flow with respect to the capillary surface; hence the distribution of $D_L/\dot{Q}_c$ ratios is by nature heterogeneous over the gas exchange surface. Although $D_L/\dot{Q}_c$ distribution improves upon exercise, complete uniformity is highly improbable under physiologic conditions. In addition, the number and distribution of erythrocytes within the capillary are important factors regulating diffusing capacity. Because of their particulate nature, each erythrocyte utilizes only a small area of capillary membrane in its immediate vicinity for $O_2$ uptake. Depending on the spacing between erythrocytes (that is, hematocrit), at any given instant a fraction of the total capillary surface may be ineffective for gas transfer. The distributions of erythrocytes and their flow are most heterogeneous at rest, resulting in a low efficiency of diffusive gas uptake. This is not a problem functionally at rest because $O_2$ demand is low. During exercise more capillary pathways are perfused and the distribution of erythrocytes becomes more homogeneous so $DL$ increases. However, if blood flow increases too fast, erythrocyte transit time may become too short for diffusive equilibrium to be established, and end-capillary $O_2$ saturation will decline. Erythrocytes must also deform from biconcave disks to parachute and other irregular shapes with increasing flow; this deformation reduces their flow resistance, facilitates their transit through the capillaries, and improves their distribution among capillary segments. However, irregularly shaped erythrocytes are less efficient in diffusive gas uptake than disk-shaped cells since a part of the erythrocyte membrane becomes hidden from the alveolar–capillary surface. These are all problems that may potentially limit capillary $O_2$ uptake; how important they are needs to be studied further.

In summary, ventilation, perfusion, and diffusion are intricately linked processes in the lung, and the pulmonary gas exchanger is designed such that it seeks the optimal balance among these processes. The notion of an "excess" capacity is misleading, since the large anatomic alveolar–capillary surface may not be completely available for gas exchange during exercise due to inhomogeneous regional distributions of ventilation, diffusion, perfusion, and erythrocytes. Because of the complexities of these interactions, uniform $DL/\dot{Q}_c$ and $\dot{V}_A/\dot{Q}_c$ distributions are not possible even in normal lungs. Hence functionally there is no excess capacity; even in normal individuals a mild diffusion limitation could develop during heavy exercise. This limitation is exaggerated by exposure to high altitude and after lung resection, but in each case the animal is able to compensate through a variety of respiratory and non-respiratory mechanisms.

## Acknowledgments

Supported by National Heart, Lung and Blood Institute grants HL40700, HL45716 and HL46185, and the American Heart Association Established Investigator Award.

## Further reading

Hsia, C. C. W. and Johnson, R. L., Jr. (1992) Exercise physiology and lung disease. In *Comprehensive Textbook of Pulmonary and Critical Care Medicine*, vol. 1, section B-5, pp. 1–20. Ed. R. Bone. Mosby Yearbook Inc., St Louis.
Hsia, C. C. W. and Johnson, R. L., Jr. (1997) Physiology and morphology of postpneumonectomy compensation. In *The Lung. Scientific Foundations*,

2nd edn, vol. 1, pp. 1047–59. Eds. R. G. Crystal, J. B. West, E. R. Weibel and P. J. Barnes. Lippincott-Raven Press, Philadelphia and New York.

Johnson, R. L., Jr., Heigenhauser, G. J. F., Hsia, C. C. W., Jones, N. L. and Wagner, P. D. (1996) Determinants of gas exchange and acid-base balance during exercise. In *Handbook of Physiology*, Section 12: *Exercise Regulation and Integration of Multiple Systems,* vol. 1, pp. 515–84. Eds. R. L. B. and J. T. Shepherd. American Physiological Society and Oxford University Press, New York.

Piiper, J. (1994) Search for diffusion limitation in pulmonary gas exchange. In *The Pulmonary Circulation and Gas Exchange,* vol. 1, pp. 125–45. Eds. W. W. Wagner, Jr. and E. K. Weir. Futura, Armonk, NY.

Scheid, P. and Piiper, J. (1997) Diffusion. In *The Lung. Scientific Foundations,* 2nd edn, vol. 2, pp. 1681–92. Eds. R. G. Crystal, J. B. West, E. R. Weibel and P. J. Barnes. Lippincott-Raven Press, Philadelphia and New York.

Weibel, E. R. (1994) Exploring the structural basis for pulmonary gas exchange. In *The Pulmonary Circulation and Gas Exchange*, pp. 19–45. Eds. W. W. Wagner, Jr. and E. K. Weir. Futura, Armonk, NY.

## 7.3 The lungs of the flying vertebrates – birds and bats: is their structure optimized for this elite mode of locomotion?

JOHN N. MAINA

The insects, pterodactyls (now extinct), birds, and bats are distinctively the only groups of animals that have ever achieved powered flight – chronologically in that order. The study of volant animals offers a particularly good example for evaluation of the evolution of complex adaptations. Flight as a means of locomotion had extensive influence on the form and function of the animals which achieved it and imposed not only development of high aerobic capacity but also called for fundamental modifications, integration, and commitment of practically all organ-systems – especially the sensory, locomotory, and respiratory systems – to this single activity. Among vertebrates, flight has only evolved in two groups: the birds and bats. Consequent to their different phylogenetic backgrounds, it is likely that different genetic resources, designs, and strategies have been utilized by these groups to independently meet the aerodynamic and energetic demands for flight. The distinct schemes that were employed presumably reflect the singular limitations of design, evolutionary adaptations to particular selective regimes, the changing physiological conditions and the developmental processes and states the animals have encountered. Considering the high energy demand of flight, comparative study of bats with birds and non-flying mammals (NFMs) is therefore a good example for working out the mechanisms of convergence in structural and functional systems in adapting to special demands. Since birds possess distinct features which do not occur in bats, it is of interest to attempt to identify the congruencies and the disparities in the individual solutions to the common energetic requirements and the commensurate ecological and physiological adjustments to locomotory needs in the two taxa.

While substantial data on the biophysical, biochemical, structural, and physiological adaptations of the avian and chiropteran flight are now

177

available, quantitative data on the pulmonary system, a significant entity in the evolution of flight, are relatively scarce, less detailed and rather narrow in scope and extent. Based on comparative physiological and morphometric data from a wide taxonomic range of birds, bats, and NFMs of different phylogenetic backgrounds, body sizes and exercise capabilities, this account outlines the high degree of integration of the parameters and factors involved in $O_2$ transfer, uptake, and transport; means by which in birds and bats gas exchange efficiency has been enhanced to support flight, a highly aerobic mode of locomotion.

## Structural adaptations of the avian and chiropteran lungs

### *Birds*

The avian respiratory apparatus (RA), the lung–air-sac system, consists of a small compact lung which is connected to capacious air-sacs. In a synchronized bellows action, the sacs ventilate the lung in a continuous and unidirectional manner. Gas exchange occurs both across a cross-current arrangement between the parabronchial air and the blood capillaries (BCs), and across a counter-current arrangement between the BCs and the air capillaries (ACs). The volume of the avian RA on average constitutes about 20 percent of the body volume (BV), with the value in some species being as high as 34 percent whereas the volume of the lung (VL) constitutes only 2.5–3 percent of BV and 11–20 percent of that of the RA. The pulmonary capillary blood volume (PCBV) makes about 36 percent of the VL, with 58–80 percent of it in the BCs. A high respiratory surface area (RSA) per unit volume of the parenchyma is achieved by intense subdivision of the parenchyma in the form of ACs which interdigitate intimately with the BCs (Figure 7.4) maximizing the RSA. The ACs of the avian lung (AL) are 3–14 µm wide, a size which is one-tenth that of the smallest alveoli of the mammalian lung (ML), which is about 35 µm diameter in the shrew, the smallest mammal with the highest known metabolism. The surface density of the blood–gas barrier (B–GB) in the AL ranges from 172 in the domestic fowl to 389 $mm^2 mm^{-3}$ in the hummingbird; in the ML, the values are about one-tenth those of birds and in the rather smooth, saccular reptilian lungs, the values range from 1.34 to 1.80 $mm^2 mm^{-3}$. Interestingly, the intense subdivision of the exchange tissue of the AL occurs in a much smaller lung with a proportionately lower parenchymal volume density of only about 46 percent. This process has been possible owing mainly to the compact (inexpansile)

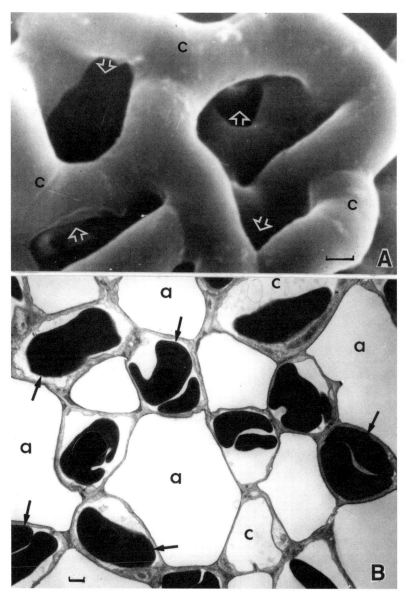

Figure 7.4. (A) Scanning electron micrograph of a latex cast preparation of the intensely branching blood capillaries (c) of the lung of the domestic fowl (*Gallus gallus* var. *domesticus*) showing the anastomoses of the blood capillaries as they intimately intertwine with the air capillaries (arrows) to maximize the surface area available for gas exchange. Though the air and blood capillaries intimately relate to each other, the air capillaries are wider than the blood capillaries and branch more profusely. (B) Transmission electron micrograph of the gas exchange tissue of the lung of the house sparrow (*Passer domesticus*) showing the air capillaries (a) and blood capillaries (c) which closely interdigitate with each other, maximizing the respiratory surface area. Note the excellent exposure of the capillary blood to air. Arrows indicate erythrocytes. Scale markers are 1 μm long.

nature of the AL which, unlike the inherently compliant ML, is ventilated by the air-sacs through a system of parabronchi that pervade it. Furthermore, the rigidity of the AL affords the remarkable thinness of the B–GB, which is in the order of 0.1 μm compared to 0.5 μm in the ML. The large PCBV and the great RSA result in a high pulmonary diffusing capacity of oxygen ($D_{L_{O_2}}$) (Figure 7.5) which is, however, somewhat comparable to that of the mammalian lungs.

In the AL, the epithelial surface area of the ACs is close to that of the capillary endothelium, with the result that the pulmonary capillary blood is intensely exposed to air. The harmonic mean thickness of the B–GB ($\tau_{ht}$) of the AL ranges from a mere 0.09 μm in the rock martin to 0.50 μm in the Humboldt penguin: the thickest B–GBs of the AL are comparable to those of the lungs of the small, energetic shrews. In birds, bats, and NFMs, $\tau_{ht}$ appears to be highly optimized. The $\tau_{ht}$ in the smallest shrew of 2.2 g is 0.230 μm, while in the huge bow-head whale, the value is only 0.35 μm. In a 7 g hummingbird, $\tau_{ht}$ is 0.1 μm compared to 0.24 μm in the

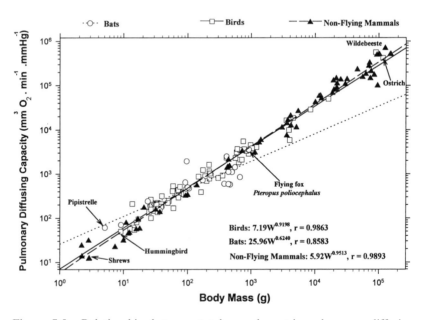

Figure 7.5. Relationship between total morphometric pulmonary diffusing capacity ($D_{L_{O_2}}$) of the non-flying mammals (NFMs), bats and birds with body mass. At a body mass below 100 g, birds have greater $D_{L_{O_2}}$ than the NFMs, while above 3 kg, NFMs have higher values. Smaller bats appear to have a higher $D_{L_{O_2}}$ than birds and NFMs.

ostrich weighing 100 kg. A pipistrelle bat (5 g) has a $\tau_{ht}$ of 0.21 µm, only little less than the 0.30 µm in a flying fox of 900 g.

In summary, the fundamental adaptive features of the avian RA include: (a) the capacious air-sacs which provide a large tidal volume and effect a through parabronchial air-flow, gas exchange occurring in the highly efficacious cross-current system; (b) large hearts which provide a high cardiac output (CO); (c) a short pulmonary circulatory time which maximizes the utilization of the hematological factors in gas uptake and transfer; and (d) favorable morphometric parameters such as a large PCBV and a particularly thin B–GB. The highly refined structural parameters result in an anatomical diffusing factor, the ratio $RSA/\tau_{ht}$, which is eight times greater than that of the NFMs of similar body size. The synergy of the physiological and anatomical factors in the AL has produced an exceptionally efficient gas exchanger with a remarkable functional reserve.

### Bats

The general anatomy of the body and the microscopic structure of the bat lung (BL) are similar to that of the NFMs. Hence, the pulmonary refinements necessary to support chiropteran flight must have been made within the constraints presented by the inferior ML. Clearly, the avian RA is not a prerequisite for flight as in bats a typical ML has been quantitatively adjusted to exchange respiratory gases (during flight) at rates equal to those of the seemingly better adapted birds. To support volancy, a repertoire of respiratory parameters has been utilized. These include $O_2$ transfer, uptake, and transport mechanisms, and encompass a relatively large CO, high hematocrit (Hct), high hemoglobin concentration (Hbc), and high blood $O_2$ carrying capacity (BO$_2$c). A venous Hct of 68 percent has been reported in a 13 g bat, *Tadarida mexicanobrasiliensis* and respectively Hbc and RBC numbers (nRBC) as high as $24.4\,\mathrm{g\,dl^{-1}}$ and $26.2$ million $\mathrm{\mu L^{-1}}$ have been encountered in some bats: this is as high as the values observed in highly trained racehorses. Whereas elevated Hct may enhance the aerobic capacity by increasing $O_2$ delivery to the tissues, the advantages thus conferred may be heavily compromised by the greater energetic demand needed to drive the more viscous blood through the vasculature: increase in the Hct beyond optimal level causes a reduction in CO, $\dot{V}_{O_2\,max}$, and aerobic scope.

Morphometrically, bats have remarkably large highly specialized lungs. The RSA $W^{-1}$ of $138\,\mathrm{cm^2\,g^{-1}}$ in the fruit bat, *Epomophorus wahl-*

*bergi*, is the highest value in a vertebrate lung and is notably greater than that of about $90 \, \text{cm}^2 \, \text{g}^{-1}$ of the hummingbird. Compared with NFMs, bats have a relatively thin B-GB, with the thinnest one in mammals ($0.1204 \, \mu\text{m}$) occurring in *Phyllostomus hastatus*, a bat with a Hct of 60 percent and an $O_2$ extraction factor ($E_{L_{O_2}}$) equal to that of a bird of comparable size. As a result of these adaptations, the $D_{L_{O_2}}$ of the smaller bats is similar to that of birds and NFMs, whereas in the large bats it seems to be lower (Figure 7.5).

## Conclusions

The convergence and/or intersection of the allometric regression lines of the pulmonary structural parameters in birds, bats, and NFMs (for example, Figure 7.5) indicates that the respiratory systems of these groups have structurally responded differently to functional demands. Changes in the allometric constants are known to occur in the course of evolution and development of animals. In the NFMs, those species which weigh between 2.5 and 260 g have considerably higher mass specific $\dot{V}_{O_2}$ than that which is expected from the entire mammalian population. Thus, small animals are physiologically not simple extensions of the larger ones. Definite adaptations to meet their greater $O_2$ demands are to be expected. Indeed, lower $O_2$ affinity of blood has been reported in such animals. Bats span a body mass range of about 1.6 g for the Thai bat, *Craeonycteris thonglongyai*, to about 1.5 kg for the flying fox, *Pteropus adulis*. The heaviest bird which can fly weighs about 18 kg and is hence about an order of magnitude heavier than the largest bat. This difference may in part be explained by the greater efficiency and functional reserve of the AL.

In general, small bats have structurally better adapted lungs than birds and NFMs of equivalent body mass, and at about 1 kg, the values of bats correspond with those of the NFMs or are even lower. The apparent deterioration of the chiropteran pulmonary parameters with increasing body mass may help to explain the poor flight performance of the larger species of bats. The high energetic demands for flight combined with limitations in the allometric scaling of the factors associated with $O_2$ uptake and transfer have greatly determined the optimal size for volancy in the group. That there are no species of bats which have given up flight altogether, as some birds have, shows the importance and the selective advantage of flight for survival in bats. Evidently, bats have exploited and probably exhausted practically all possible adaptive resources to achieve and maintain flight by variably utilizing an arsenal of functional and structural parameters.

Since the energetic cost of flight in birds and bats is essentially the same, and since the basic designs of their gas exchangers are remarkably different, comparable degrees of respiratory optimization for gas exchange to support flight must have been achieved through somewhat different stratagems. In birds, this appears to have occurred through a process where a wide range of pulmonary and extrapulmonary factors and processes were integrated into the gas exchanging machinery, with certain factors operating at suboptimal levels: we have termed this scheme of operation a "broad based–low keyed" strategy (Figure 7.6). In bats, the process seems to have taken place through a synergism of performances of a rather restricted number of factors which appear to be remarkably highly refined, thus generally operating close to their maximal level: this plan has been termed the "narrow based–high keyed" strategy. Expounding on these concepts, in bats, factors ranging from nRBC, Hbc, $BO_2c$ to pulmonary morphometric parameters are evidently all very highly tuned, while on the other hand birds have not found it necessary to enhance parameters such as the respiratory rate – RR (a RR of 3 times per minute has been reported in the swan) and blood $O_2$ uptake and transfer properties are essentially similar to those of the NFMs (Figure 7.6). Furthermore, while during flight bats utilize wing beat to help move air in and out of the lung and hence a 1:1 ratio between RR and wing beat prevails, in birds such relationships rarely occur. The capacity to utilize a large number of respiratory factors and additively stretch them in response to need and circumstances gives birds their remarkable potential to respond to a wide range of conditions rather than optimally for a particular set. This must call for a compromise of design to accommodate even conflicting parameters and conditions. In the bar-headed goose, which flies at very high altitudes, CO increases seven-fold during hypoxia, especially when the arterial $P_{O_2}$ falls below 28 torr; and in the barnacle goose, during flight, the heart rate may increase seven-fold of the resting value, and the stroke volume two-fold. Though some species are known to tolerate experimental hypoxia and hypobaria as well as birds, bats are not known to fly at altitude, a capacity birds have mastered. The different schemes for gas exchange that birds and bats have adopted have culminated in comparably efficient gas exchangers which through an integrative effect provide the necessary quantities of $O_2$ needed for flight. Inescapably, the different strategies should result in disparate functional reserves/safety margins, with the

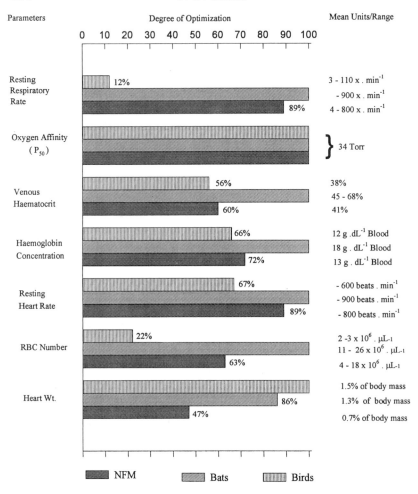

Figure 7.6. Relative degrees of optimization of some respiratory parameters in birds, bats, and non-flying mammals (NFMs), chosen solely on the basis of availability of adequate published data on the three taxa. The group with the highest value was given a score of 100 percent and the others proportionately related to it. Birds have exploited these factors to a smaller extent (average percentage of exploitation, 60.4 percent) compared with bats (98 percent) and the NFMs (74 percent); see text for discussion.

scope in bats being rather narrow. This contention is supported by the observation that whereas with changing ambient temperature birds and NFMs adjust respiratory frequency (f) and tidal volume ($Vt$) to meet changing $O_2$ demands, while $E_{LO_2}$ remains fairly untouched (and in some species actually decreases), bats accommodate varying thermogenetic $\dot{V}_{O_2}$ by simultaneously changing all the three factors, with $EL_{O_2}$ increasing by as much as two-fold. Analysis of the available data indicates that birds have exploited the pulmonary structural parameters in preference to the physiological ones, while bats appear to have "favored" the physiological ones: about 60 percent of the gas exchange capacity in birds can be attributed to the structural parameters and 40 percent to functional ones, while in bats, 61 percent is attributable to the physiological factors and 39 percent to the structural ones.

These observations affirm the holistic view that, in nature, there is no single or simple common mechanism of adapting to specific functional demands but all means and resources can in different combinations be utilized to achieve the same end results. The multidimensionality of such complex permutations make the state and process of optimization and the particular factors behind such adjustments difficult to define precisely.

## Acknowledgments

This manuscript was prepared during tenure of a CIDA/NSERC fellowship at McMaster University, Ontario, Canada. I wish to thank Drs C. M. Wood, H. L. Bergman, A. N. Bergman, Chow-Fraser, and W. Yu-Xiang for numerous courtesies.

## Further reading

Maina, J. N., King, A. S. and Settle, J. G. (1989) An allometric study of the pulmonary morphometric parameters in birds, with mammalian comparisons. *Phil. Trans. R. Soc. B, Lond.*, **326**, 1–57.

Maina, J. N., Thomas, S. P. and Hyde, D. M. (1991) A morphometric study of bats of different size: correlations between structure and function in the chiropteran lung. *Phil. Trans. R. Soc. B, Lond.*, **333**, 31–50.

Scheid, P. (1979) Mechanism of gas exchange in bird lungs. *Rev. Physiol. Biochem. Pharmacol.*, **86**, 137–86.

Schmidt-Nielsen, K. (1984) *Scaling: Why is Animal Size so Important?* Cambridge University Press, Cambridge, UK.

Thomas, S. P. (1987) The physiology of bat flight. In *Recent Advances in the Study of Bats*, pp. 75–99. Eds. M. B. Fenton, P. Racey and J. M. V. Rayner. Cambridge University Press, Cambridge, UK.

Weibel, E. R. (1984) *The Pathway for Oxygen*. Harvard University Press, Cambridge, MA.

# 7.4 Gills of water-breathers: structures with multiple functions

## EDWIN W. TAYLOR

The gills of water-breathing animals are generally regarded as respiratory organs, that is the equivalent of the lungs in air-breathers, responsible for the uptake of oxygen and the excretion of carbon dioxide. However, gills subserve many other functions largely attributable in lung-breathers to the kidneys, such as nitrogen excretion, ionoregulation and acid-base regulation. Accordingly, consideration of the extent to which the concept of symmorphosis is applicable to gills requires that this multiplicity of functions be considered both separately and in combination, because many of them are functionally connected.

The exchange of gases over the gills is predominantly by simple physical diffusion so that exchange is governed by the relative diffusion conductance of the gills for each gas. Approximately equal amounts of oxygen and carbon dioxide are transported across the gills of fish undergoing aerobic metabolism, whereas ammonia excretion rates vary between 10 and 30 percent of the rate of oxygen uptake. As carbon dioxide is 20–30 times and ammonia 200–300 times more soluble than oxygen in water, their permeation coefficients (diffusion coefficient $\times$ solubility) scaled against oxygen are 23 for $CO_2$ and 32,600 for $NH_3$. Consequently, if the partial pressure gradient across the gills is 50 torr for $O_2$ it will be about 3 torr for $CO_2$ and 16 $\mu$torr for $NH_3$ (Randall 1990). These markedly different gradients determine that while changes in gill morphology and ventilation can have a marked effect on blood oxygen levels, the effect is minor on $CO_2$ and negligible on $NH_3$ levels. The overall morphometric design of the gills, therefore, can be directly related to their function in oxygen uptake, and the control of ventilation is strongly coupled to the maintenance of the oxygen requirements of the animal.

## The gills as gas exchangers

Respiratory gas exchange over gill lamellae takes place across a functional countercurrent, with blood flow in the lamellae running counter to water flow between them. This has clear advantages both in terms of the partial pressures achieved in arterialized blood, which can be higher than in expired water, and in determining the effectiveness of oxygen uptake.

Model analysis of gas transfer in fish gills, which compares them with the mammalian lung, is typically confined to a consideration of the factors governing oxygen uptake (e.g. Piiper and Scheid 1982). This analysis has concluded that the countercurrent system operating across the exchange surfaces on the gill lamellae provides the potential for the highest possible effectiveness of gas exchange, in that complete equilibration of arterial blood to inspired water, and of expired water to venous blood, is theoretically possible. In reality, complete equilibrium is obstructed by diffusion resistance, mismatch between water and blood flows, and both water and blood shunts (Figure 7.7). Nevertheless, a routinely active trout can extract as much as 80 percent of the available oxygen from the respiratory stream.

Symmorphosis seems to imply that the structural and functional characteristics of a system are genetically determined, so that a morphometric study can reveal meaningful relations between, for instance, oxygen requirements and the calculated diffusion conductance of the gills. This has revealed clear relationships between the surface area of the gill lamellae, the harmonic mean of the diffusion distances between body fluids and water, and the relative oxygen requirements of the fish. These vary with body size, and with lifestyle to the extent that it affects metabolic capacity. Maximum diffusivity is characteristic of highly active species such as tuna, and minimum levels with lethargic species such as the toadfish. Gill area varies by a factor of over ten between skipjack tuna and toadfish when body mass is scaled to 1 kg (Hughes 1989). Of equal importance is diffusion distance, and this is found to vary inversely with oxygen demand, such that tuna have relatively large gill surface areas and relatively thin diffusion barriers, while inactive species such as the dogfish have relatively smaller gill areas and longer diffusion pathways. These characteristics show convergence with the lamellar gills of crabs, implying that variations in gill morphometry in relation to oxygen demand are ubiquitous.

There is a consistent anomaly with respect to morphometric measurements on fish gills, in that calculated diffusing capacity for oxygen ($D_{O_2}$)

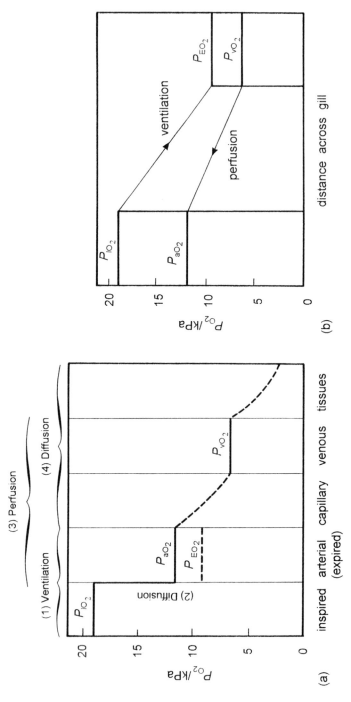

Figure 7.7. (a) Oxygen cascade of mean $P_{O_2}$ values for aerated water to tissues in a typical fish. $Pa_{O_2}$ is greatly reduced relative to $P_{I_{O_2}}$ because the gills present a considerable barrier to diffusion. However, $Pa_{O_2}$ is often higher than $P_{E_{O_2}}$, due to the counter-current of blood and water at the gill lamellae. The oxygen cascade consists of both convective and diffusional oxygen transfer. (b) Diagram of counter-current showing how $Pa_{O_2}$ may exceed $P_{E_{O_2}}$.

is always much higher than that determined by physiological measurements. The reasons for this reside in the complex dynamics of the system. Water and blood flow through the gills are to some extent laminar, which leaves virtually unstirred layers at the exchange surface. Because of the relatively slow rate of diffusion of oxygen through water, these layers make a large contribution to the total diffusion barrier at the gills. In addition, a proportion of the water entering the gill sieve is shunted past the lamellae and does not effectively contribute to gas exchange.

Several other factors complicate the relationship between gill morphometrics and oxygen demand. Some oxygen is taken up over the general body surface. This can be particularly important in embryo, larval, and juvenile fish which are small in size, yielding a high surface area to volume ratio for exchange, and have incompletely developed gills. The surface of structures such as the yolk sac, which is relatively well perfused, may be important in gas exchange. In adult fish the general body surface is relatively impermeable and poorly perfused and ventilated compared with the gills. Nevertheless, about 10–12 percent of total oxygen uptake can occur directly over surface structures other than the gills, satisfying local oxygen demand, particularly during hypoxia, rather than providing oxygen to the circulation.

Another factor complicating the relationship between gill morphometrics and oxygen demand is that patterns of perfusion and ventilation of gill lamellae may vary. Effective surface area for gas exchange can be affected by patterns of perfusion; for example, in an inactive trout only about 60 percent of gill lamellae are perfused. This proportion rises during active swimming and hypoxia and the consequent increase in the area available for gas exchange leads to an increase in transfer factor – a measure of the diffusive conductance of the gills. Clearly there are considerable safety factors built into the branchial exchanger.

### The gills in ion exchange

So far we have only considered the functioning of the gills in oxygen uptake, because this largely determines their gross dimensions. However, as stated earlier, the diffusivity of the gills for the other gases they exchange ($CO_2$ and $NH_3$) is much greater than for oxygen, so that virtual equilibrium between blood and water levels is achieved in a fraction of the diffusion time required for oxygen. In terms of these exchanges, the relative flow rates of water and blood, which are regulated in order to optimize oxygen uptake, are excessive (that is, they contain large safety factors). Surprisingly, however, the trout shows ventilatory changes in response to bicarbonate and ammonia infusion. The former effect is attributable to the reduction in pH following an increase in $CO_2$ which reduces blood oxygen content in the short term due to the Root effect, a phenomenon to be discussed in Chapter 7.5. The effect of ammonia is probably direct and relates to it being highly toxic.

The processes of ion transport and acid-base regulation over fish gills are still not fully understood, and in particular the division of function between pavement epithelia and mitochondria-rich cells (MRC) is unclear. However, a working model of exchange is provided in Figure 7.8. Various ions are actively exchanged at the gills, either by ATP-depen-

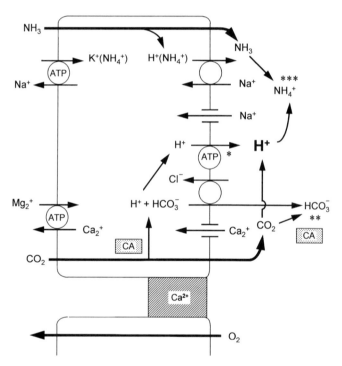

Figure 7.8. Schematic diagram of gas and ion exchange over the gill epithelium of a fish. Oxygen, carbon dioxide, and ammonia predominantly diffuse as gases. Some carbon dioxide is hydrated in the epithelial cells to yield protons and bicarbonate, while ammonia yields ammonium ions which may be exchanged for sodium or chloride ions at the apical membrane. The gill boundary layer is acidified, either by an electrogenic proton pump (*), or by hydration of carbon dioxide catalyzed by carbonic anhydrase (CA**). This results in ammonia being trapped as nonpermeating ammonium ions in the water (***), sustaining the diffusion gradient for ammonia from blood to water.

dent pumps or via ion channels. These processes can be rate-limited by
the gradients for their exchange, and by the availability of exchange sites;
that is, they show saturation kinetics. Despite the intervention of these
active ionoregulatory processes, which enable gill-breathers to accumu-
late or eliminate ions against concentration gradients, ion and water
fluxes can change with oxygen transfer. For example, an increase in the
oxygen transfer factor in the freshwater trout during exercise is associated
with an increased rate of sodium exchange and also a net flux of water
into the body across the gills. This example of an apparent conflict
between functions of the gills has to be countered by increased rates of
urine production.

The principles of symmorphosis can be applied to the processes of
branchial ionoregulation. What is required is a measure of the active
areas available for ion exchange, plus some measure of their relative
effectiveness. Depletion of major ions in the environment (for example,
$Na^+$ and $Ca^{2+}$) stimulates proliferation and increased exposure (from
pits in the epithelium) of MRCs, leading to an increase in the effective
surface area and a consequent increase in the rate of flux of the ions.
Similar changes happen on the gills and in the gut of fish transferred from
freshwater into seawater, when they are potentially flooded with ions they
have to eliminate. Thus, a simple count of MRC numbers in thin sections
of fish gills could be taken as an index of their response to a change in the
ionoregulatory demand. However, this analysis is complicated at present
by our lack of a complete understanding of the processes involved and of
their sites of action.

## Perturbation of the system by toxic damage

The gills are particularly susceptible to the presence of toxic chemicals
dissolved in the water. Many of these affect the exchange properties of
the gills, and an analysis of their effects may illuminate the consideration
of symmorphosis. Exposure of freshwater rainbow trout to lethal concentrations of copper at acid pH causes recognizable damage to the gill
lamellae. Epithelial cells slough off, blood spaces coalesce, and a mixture
of cell debris and mucous accumulates on the surface. Death follows after
about 24 h, but is preceded by a succession of functional changes.
Ionoregulation breaks down so that sodium and chloride levels drop.
The animal becomes progressively hypoxic, then hypercapnic.
Accumulation of lactic acid, together with carbon dioxide, causes a developing acidosis. As part of an attempted adaptational response to these

changes the fish increases ventilation rate and releases catecholamines into the blood stream. This in turn causes a release of red cells from the spleen so that haematocrit may rise to 50 percent (that is, three times normal). Blood protein levels rise, indicating a reduced blood volume. Heart rate and blood pressure rise and the fish finally appears to succumb to heart failure. What is of interest here is that failure of oxygen uptake is secondary to ionoregulatory problems. The immediate attempts the fish makes to counter acute pollution seem directed at the problem of reduced oxygen uptake over damaged gills (that is, increased ventilation, heart rate, blood pressure, and haematocrit), due to stimulation of oxygen receptors. They are in the event inappropriate and result in cardiovascular collapse.

The effects of sublethal levels of copper reveal some of the problems intrinsic in the application of the concept of symmorphosis to complex systems such as fish gills (Taylor *et al.* 1996). Trout exposed to levels of copper that were just below the incipient lethal level ($0.5 \, \mu mol \, l^{-1} \, Cu^{2+}$ in freshwater at pH 5 and 5 °C) survived exposure but were unable to maintain sustained (aerobic) swimming. Sublethal copper exposure caused damage to gill ultrastructure, including hyperplasia of epithelial cells, proliferation of mucocytes, necrosis of MRCs, and lamellar fusion. Morphometric analysis showed the harmonic mean diffusion distance to have tripled. On these grounds it is a reasonable assumption that the fish are unable to take up sufficient oxygen to supply aerobic exercise. However, a physiological study demonstrated this was not the case. Routine levels of oxygen uptake increased following copper exposure and, despite extensive gill damage, the fish were not hypoxic, though arterial $P_{O_2}$ fell during exercise, indicating a diffusional limitation on the gills. Maintenance of arterial $P_{O_2}$ may have resulted partly from a 40 percent increase in gill ventilation rate. Oxygen content of the blood was also maintained partly by an increase in haematocrit. While there was no increase in plasma lactate in trout exposed to sublethal copper, lactate levels in red muscle doubled, suggesting that haemoconcentration (evidenced by increased plasma protein levels) may disrupt oxygen transport in the periphery, causing local tissue hypoxia.

More surprisingly, exposure to copper, and to acid water alone, caused a progressive increase in plasma ammonia levels. This effect cannot be due to changes in gill diffusivity because the fish were not hypoxic and the acid water should enhance ammonia trapping (Figure 7.8), thus maintaining an effective diffusion gradient for ammonia excretion. These data imply that active secretion of ammonium ions, possibly in exchange for sodium, is an important component of ammonia excretion, which is disrupted by copper. Carbon dioxide excretion is known to entail some active secretion of protons and bicarbonate (Figure 7.8), though the mechanisms remain unresolved. Thus, the exchange of carbon dioxide and ammonia over the gills may not relate directly to their relative diffusivity, and in turn to gill morphometrics. Reduction in swimming speed following copper exposure was in direct proportion to levels of ammonia accumulation, possibly because ammonia affects nervous transmission or integration by displacing potassium ions, or may fatigue muscle by depletion of the intermediates or products of oxidation metabolism. Whatever the causality, it is clear that oxygen supply and the morphometric considerations that govern its supply are not the only factors limiting the scope for aerobic exercise in fish.

## Conclusion

The dominant function of fish gills, which determines their gross morphometrics and their rates of ventilation and perfusion, is the uptake of oxygen. The relationship between gill morphometrics and oxygen uptake is complicated by ventilation–perfusion mismatching and by shunting. In addition, factors affecting the roles of the gills in ionoregulation and excretion may limit the performance of the fish. Some of the mechanisms illustrated in Figure 7.8 are common to the exchange epithelia of all animals, and have been described both in fish gills and the mammalian kidney. Whether they be considered as ancestral characters, or to have resulted from convergence, they are likely to be of considerable importance in any analysis of gill function based on the concept of symmorphosis.

## Further reading

Hughes, G. M. (1989) Morphometry of respiratory systems. In *Techniques in Comparative Respiratory Physiology*, pp. 21–48. Eds. C. R. Bridges and P. J. Butler. SEB Seminar Series, 33. Cambridge University Press, Cambridge.

Piiper, J. and Scheid, P. (1982) Physical principles of respiratory gas exchange in fish gills. In *Gills*, pp. 45–62. Eds. D. F. Houlihan, J. C. Rankin and T. J. Shuttleworth. SEB Seminar Series, 16. Cambridge University Press, Cambridge.

Randall, D. J. (1990) Control and co-ordination of gas exchange in water breathers. In *Advances in Comparative and Environmental Physiology*, vol. 6, pp. 253–78. Ed. R. G. Boutilier. Springer-Verlag, Berlin.

Taylor, E. W., Beaumont, M. W., Butler, P. J., Mair, J. and Mujallid, M. S. I. (1996) Lethal and sub-lethal effects of copper upon fish: a role for ammonia toxicity? In *Toxicology of Aquatic Pollution*, pp. 85–113. Ed. E. W. Taylor. SEB Seminar Series, 57. Cambridge University Press, Cambridge.

# 7.5 Factors influencing the optimization of hemoglobin oxygen transport in fish

## DAVID RANDALL

A historical tendency in animal physiology has been to treat each of the regulated physiological systems of animals more or less separately, with relatively little focus on their mutual functional and structural interdependencies. However, the actual design and function of an individual physiological system is modified by constraints placed on it because it is part of a larger physiological network. Because these systems are mutually dependent, environmental stresses may present conflicting demands upon individual systems. The concept of symmorphosis is that the capacities of component parts of a physiological cascade, such as the transfer of oxygen from the lungs to the tissues, are matched. A component part, however, may belong to several systems, and the capacity of the component may be constrained by the requirements of other linked systems.

### Hemoglobin structure and O₂ delivery

Consider the following example. The blood of Teleost fish shows a so-called Root shift, which refers to the fact that a drop in blood pH reduces the ability of the respiratory pigment hemoglobin to bind oxygen. As a consequence, a reduction in blood pH as it occurs in the swimbladder due to acidification of blood in the gas gland will reduce the oxygen carrying capacity of the blood; oxygen is released from the hemoglobin and this will raise the partial pressure of oxygen in solution in the blood, resulting in oxygen diffusion into the swimbladder. Thus oxygen released from the blood is used to inflate the swimbladder. This helps keep the fish neutrally buoyant and reduces the amount of energy that must be expended on maintaining position in the water column.

A conflict between physiological systems emerges, however, when a fish is forced into strenuous exercise, and accumulation of lactate causes a general drop in blood pH. The effects of this acidosis could be expected to reduce blood oxygen capacity and, therefore, interfere with oxygen delivery by hemoglobin to all tissues. This would, in turn, impair the aerobic swimming capacity of the fish. However, this potential conflict between oxygen transport and the need for activity is resolved by the secretion of large quantities of catecholamines following strenuous exercise. The catecholamines stimulate Na/H exchange in red blood cell membranes and thus raise erythrocytic pH, even as plasma pH remains low due to the presence of metabolic acids in the blood. The high erythrocytic pH prevents the plasma acidosis from generating a Root shift in the blood, and so oxygen transport to tissues is not seriously affected by the acidosis associated with strenuous exercise.

In solving one potential conflict, however, another conflict is generated! An acidosis in the gas gland cannot generate a Root shift. Following strenuous exercise the fish maintains oxygen delivery to the tissues, but it does so at the expense of oxygen secretion into the swimbladder. Interestingly, catecholamines not only raise erythrocytic pH, but also reduce blood flow to the gas gland and, therefore, gas secretion into the swimbladder. When acid–base disturbances are eventually corrected, catecholamine levels fall and the fish can once again use a local acidosis in the gas gland to secrete oxygen into the swimbladder. Thus the nature of the cascade and the capacity of the component parts (in this example the functional capacity of the Root shift) changes in time, depending on the physiological situation.

### Species differences in the gas transfer cascade

Component parts of different animals serving similar functions may have different properties depending on the extent of linkage with other systems. For example, trout hemoglobin has a marked Root shift but a low buffering capacity, whereas hemoglobin from dogfish has no Root shift but a large buffering capacity. The difference in buffering capacity is related to the number of histidine residues which is much lower in trout hemoglobin than in that of dogfish (Figure 7.9). Both of these water-breathing fish are strong swimmers, capable of prolonged periods of aerobic swimming, so differences in exercise capacity and, therefore, oxygen and carbon dioxide flux through the system, are unlikely to be associated with the differences in hemoglobin structure. The major difference between these two fish is that the trout can live in waters of very different salt content while the dogfish is limited to sea water. How then does the nature of the aquatic environment impinge on the properties of hemoglobin? This has to do with the structure of the gills and their effect on gas and ion exchange.

### Gill structure and hemoglobin function in the trout

The gill epithelium consists of pavement, mucus, and chloride cells which serve different exchange functions. In the trout, the pavement cells, which make up most of the gill epithelium, have an apical proton ATPase associated with an amiloride-sensitive sodium channel (Figure 7.9). The proton ATPase appears to energize sodium uptake, which is then removed from the cell by a basal Na/K-ATPase. The chloride cells are fewer in number and have an apical chloride/bicarbonate exchanger.

The pavement cell has an apical distribution of carbonic anhydrase that parallels the proton ATPase activity and serves to supply protons to the apical proton ATPase. Carbon dioxide enters the epithelium from the blood and is hydrated, the proton being excreted across the apical membrane and the bicarbonate possibly being excreted across the baso-lateral membrane, presumably in exchange for chloride. The absence of carbonic anhydrase in the basal regions of the cell and in the membrane facing the plasma will slow the transfer of protons between the plasma and the gill epithelium. The absence of carbonic anhydrase from the plasma in trout results in little or no bicarbonate dehydration within the plasma because the reaction is uncatalyzed and the transit time for blood through the gill is only about a second. Thus as the blood perfuses the gills, molecular carbon dioxide diffuses across the epithelium and plasma bicarbonate enters the red blood cell; here it is dehydrated at the catalyzed rate in the presence of high intracellular concentrations of carbonic anhydrase and excreted as carbon dioxide. Because carbon dioxide and bicarbonate enter and leave the plasma by different routes and the reaction in the plasma is uncatalyzed, the carbon dioxide/bicarbonate reaction almost never reaches equilibrium in trout plasma.

### Coupling of oxygen and carbon dioxide transfer in the trout

The absence of available carbonic anhydrase in trout plasma ensures that nearly all carbon dioxide excretion is associated with erythrocytic bicarbonate dehydration, the protons being supplied by erythrocytic buffers

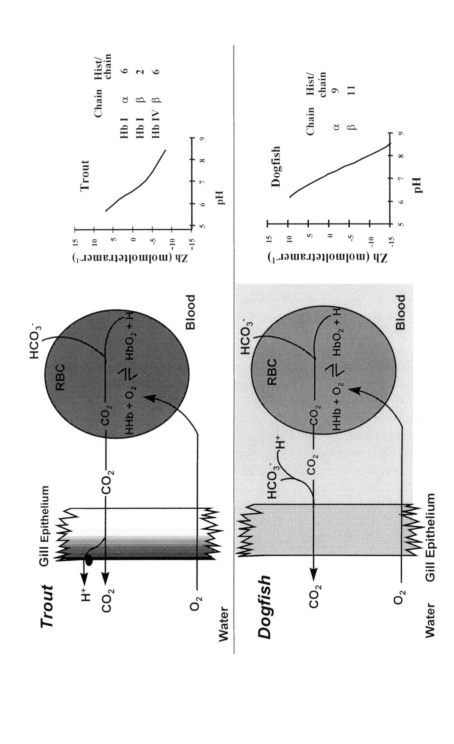

and by oxygenation of the hemoglobin. Under most conditions hemoglobin oxygenation is the major source of protons for bicarbonate dehydration. The low buffering capacity of the hemoglobin ensures that protons produced by oxygenation are available for bicarbonate dehydration. As a result, oxygen and carbon dioxide transfer is tightly coupled in trout. This tight coupling of carbon dioxide excretion to oxygen uptake in the red blood cell also ensures that protons produced by hemoglobin oxygenation are consumed by carbon dioxide excretion, ameliorating erythrocytic pH changes. In the absence of carbon dioxide flux through the red blood cell, oxygenation of the erythrocyte will reduce pH and prevent saturation of the hemoglobin due to the Root shift. Bicarbonate flux through the red blood cell, combined with catecholamine regulation of red cell pH, ensures that the Root shift will not normally limit blood oxygen content during the process of oxygenation.

### Carbon dioxide transfer in dogfish

Dogfish have less proton ATPase in the gill epithelium than trout and it is limited to a few mitochondria-rich cells in the interlamellar space and on the lamellae. These sparsely distributed cells are presumably involved in acid excretion and not in sodium or chloride uptake across the gill. The distribution of carbonic anhydrase in dogfish gills is extensive with intense labeling in some cells, possibly the acid-excreting cells. Carbonic anhydrase is also found on the surface of the gill epithelium facing the plasma, as well as in the plasma itself (Figure 7.9). This very general distribution indicates that carbonic anhydrase must play a more extensive role than simply augmenting the supply of protons to the proton ATPase, as was noted for the trout. Partial pressures of carbon

Figure 7.9.   Trout and dogfish hemoglobins are associated with different types of gill epithelial cell. The trout gill epithelium has a high proton ATPase activity associated with carbonic anhydrase activity in the apical membrane, but none on the basal–lateral membrane or available to the plasma. Trout hemoglobin has a lower buffering capacity than dogfish hemoglobin. The dogfish gill, on the other hand, has little proton ATPase activity but carbonic anhydrase activity in the gill epithelium, plasma, and red blood cell. Dogfish hemoglobin has a high buffering capacity. The shading indicates the distribution of carbonic anhydrase: the darker the shading, the greater the carbonic anhydrase activity. The graphs on the right are the $H^+$ titration curves, Zh (net $H^+$ charge, mol $H^+$ mol tetramer$^{-1}$) as a function of pH of hemoglobin from trout and dogfish (from Jensen 1989). The number of histidine side chains for several $\alpha$ and $\beta$ chains of hemoglobin from trout and dogfish is indicated to the right of each graph.

dioxide are much lower in dogfish than in trout and the extensive distribution of carbonic anhydrase could facilitate diffusion of carbon dioxide across the gills. Bicarbonate dehydration in the plasma, as well as the red blood cell, undoubtedly contributes to carbon dioxide excretion. Carbonic anhydrase activity in dogfish red blood cells is less than that in trout erythrocytes. In addition, there are no carbon dioxide/bicarbonate disequilibria in the plasma flowing away from the gills of dogfish. Carbon dioxide and oxygen transfer are not tightly coupled, and carbon dioxide can be excreted directly from the plasma. The Haldane effect is small in dogfish blood but hemoglobin buffering capacity is high, minimizing red blood cell pH changes in a very different way from that of trout. There is no Root shift nor any regulation of red cell pH by catecholamines in dogfish as seen in trout.

## Conclusions

The functional capacity of the Root shift in trout varies with the physiological state of the animal and variations in the pattern of ion transfer in trout and dogfish are associated with differences in hemoglobin structure and function in these animals because of the extensive linkage between ion and gas transfer (Figure 7.9). Thus physiological systems are constrained both in design and capacity by the requirements of other linked systems in time and space, and cannot be viewed in isolation. Given that the level of integration varies, the extent and nature of such constraints needs to be assessed in any analysis of optimization of animal design. Alternatively, symmorphosis could be used as a means to estimate the degree of linkage between systems. If the capacity of a component part is adjusted to the overall requirements of a particular system then there is probably little linkage between that component and other functional systems. Conversely, if the component part is not adjusted to the overall requirements of a particular system then there is probably extensive linkage of that component to other systems. The concept of symmorphosis may be useful, not only to analyse rate-limiting steps in a cascade, but in addition could be used as a means to determine the extent of linkage between physiological systems.

## Further reading

Jensen, F. B. (1989) Hydrogen ion equilibria in fish hemoglobins. *J. Exp. Biol.*, **143**, 225–34.

Larsen, E. H. (1991) Chloride transport by high-resistance heterocellular epithelia. *Physiol. Rev.*, **71**, 235–83.

Lin, H. and Randall, D. J. (1995) Proton pumps in fish gills. In *Fish Physiology: Cellular and Molecular Approaches to Fish Ionic Regulation*, vol. 14, pp. 229–55. Eds. C. M. Wood and T. J. Shuttleworth. Academic Press, New York.

# 8

# Nutrient supply system

## 8.1 Overview

### AMIRAM SHKOLNIK

The major function of the nutrient supply system is to provide quantitatively adequate amounts of qualitatively appropriate substrates at a rate sufficient to sustain energy-releasing oxidative reactions of the cells and accounting for the variability of the resources in nature. Since the digestive tract performs a chain of consecutive steps, optimal function of the system also implies that any single step in the chain should not become a rate-limiting factor to the entire process, not even when metabolism is subjected to high loads.

The principle of symmorphosis implies that the efficiency of the animal's digestive system which is designed to absorb from the food the substrates needed by the cells should accommodate the demands of the animals in nutrients and energy. A test of this contention was carried out in mice under peak energy and nutrient demand by subjecting them to a combination of lactation and low environmental temperatures. This showed that, even when the rate of energy expenditure is increased to twice the maintenance requirements, the capacities to eat, digest, and transport nutrients across the intestinal wall was not exhausted. A close match between energy input, enzymatic digestion, and nutrient absorption was also demonstrated in animals other than mice, suggesting that the limit to energy expenditure of the whole animal is likely to reside in the energy output pathways rather than in the input pathways of digestion and intestinal uptake capacities. The established fact that dairy cows, sheep, and goats increase their expenditure during peak lactation to four to eight times their maintenance levels also supports this conclusion.

Due to its location at the interface to the environment, the digestive system must also adapt qualitatively to allow a great variety of foods to be utilized. In that sense, the optimization of the nutrient supply system

requires further testing in herbivorous animals, especially in the species
that derive their energy from high-fiber plant material. While cellular
food constituents provide animals with readily digestible compounds
such as sugar, starch, fat, and proteins, the bulk of photosynthetic energy
that is stored in terrestrial plant material resides in structural elements of
the plant cell wall, mostly in cellulose, the most abundant carbohydrate
on earth. Most plant cell walls are made of polymers of glucose that are
not accessible to digestion by the enzymes of the digestive tract of vertebrates. Their processing depends on fermentation by symbiotic microor-
ganisms, and this requires anatomic specializations of certain segments of
the digestive system to serve as spacious vats in which the rate of passage
of digesta is slowed down sufficiently to allow for efficient microbial
fermentation. In other words, time-consuming steps must be added to
the digestive sequence. Modeling the flow of digesta through the fermentation chambers and the gut, based on the various designs of chemical
reactors, is likely to be of great help in analyzing the rate of the functions
taking place in the complicated digestive tract, particularly in herbivorous animals.

The share of plant cell wall constituents in a vegetarian diet varies
greatly depending on the type of vegetation available and on the parts
of the plant consumed; it is further affected by seasonal differences. The
specific food preference of an animal explains the wide array of different
morpho-physiological features found in the herbivorous animals' digestive tract, most notably in ruminants. A classification of herbivorous
animals according to their diet and an identification of the relationship
between the animal and its environmental food, including the mode of
food acquisition, are consequently prerequisites to the analysis of anatomical and functional optimizations in the nutrient supply system.

An aspect that is often overlooked in the discussions on optimization
of the nutrient supply system of animals is related to the changes, during
ontogenetic development, in food preference, metabolic components, and
allocation of nutrients to target organs. This was brought into focus by a
study on the ways energy metabolism changes in fish with age and ontogenetic phase of development.

# 8.2 The match between load and capacity during lactation: where is the limit to energy expenditure?

## KIMBERLY A. HAMMOND

It is an unavoidable fact of nature that animals are limited in the amount of energy they can expend in a day. Restrictions on energy expenditure can result from limited availability of nutrients in the environment, or from physiological limits on the ability to utilize those nutrients even if they are available in unlimited supply. It is this limit on physiological capacity to digest, metabolize, and expend ingested energy that I address here. Are animals designed so that their capacity to expend energy is exactly matched to their energy demand? Does that capacity change when energy demand changes?

### Proximate and ultimate limits to energy transfer

After food is ingested, it must be broken down and digested, and the resulting breakdown products must be transported across the intestinal wall to the bloodstream for further use. Because these independent physiological processes occur in series, a restriction on one portion of this transfer chain will restrict the rates of other processes that follow. Alternatively, each physiological process may be restricted in a similar manner with respect to the others so that no single process limits the overall transfer of energy; instead, there would be a coadjusted limit to energy transfer and the system would operate in an optimized manner. This is what is meant by symmorphosis.

Perhaps the best way to determine how whole-animal energy expenditure is limited by individual physiological processes is to understand how these processes are limited under the highest forms of energy demand that an animal encounters in its lifetime. These peak demands vary hourly such as during sprint running or extreme cold exposure, or on a sustained basis (existing for days or weeks), such as during migration, acclimation to cold winter environments, or, for mammals, during lactation. In general, high short-term increases in energy demand are fueled by metabolism of stored materials (for example, body fat), while relatively lower sustained energy increases are balanced by food intake. It is the food-balanced and sustained high energy demands that I examine here to understand limits to whole-animal energy transport and expenditure. For most of the studies I describe, lactation was used as a way to increase energy demands in female mice.

To understand if one or all input processes contribute to the limit on whole-animal energy expenditure, it is asked if lactational performance of the mother mouse (determined by the number and mass of pups and the mass of the litter weaned) is limited by her ability to eat, digest, and transport enough energy and nutrients across her intestine (intestinal capacity). Second, it is asked if input processes limit whole-animal energy expenditure, or if whole-animal energy expenditure is instead limited by output processes such as the mother's ability to make enough milk for her litter (Figure 8.1).

### Limitations on energy input during lactation

Is whole-animal energy expenditure determined by intestinal capacity? To answer this question, mice were placed in several different conditions. First, lactating mothers were allowed to nurse pups to the day of peaklactation (day 15 for Swiss-Webster mice) (Hammond and Diamond 1992). Three litter sizes were used to gradually increase the demand on lactating females: small litters (five pups), medium litters (eight pups),

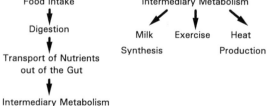

Figure 8.1. General model for experimental design.

and large litters (fourteen pups). Second, mice were acclimated the cold-est temperatures they could withstand (Konarzewski and Diamond 1994) to understand whether limitations to input processes are different for different types of energy demand. Third, a combination of moderately cold temperatures and lactation (Hammond *et al.* 1994) was used to determine if the limits observed with either cold or lactation would be the same with the combination of cold and lactation.

Animals were given a 55 percent sucrose diet. Daily food intake, small intestinal capacity for carbohydrate breakdown, and the subsequent capacity of the small intestine to transport the glucose from the lumen of the small intestine to the blood (summed uptake capacity) were mea-sured. By comparing the summed glucose uptake capacity with the amount of glucose ingested in a day, it was determined whether it was possible to absorb all of the glucose resulting from that day's food intake, and if there was an excess capacity available for glucose transport. Any excess or surplus in glucose uptake capacity would indicate that the small intestine is not limited in its capacity to transport nutrients.

During normal lactation, mothers with large litters ate more food than those with small litters. The total mass of litters weaned was greater in large litters than in small litters; however, mean pup mass from large litters was less than that from small litters. This indicates that mothers with large litters could not fully compensate for that challenge because they could not wean pups that were as large as pups from mothers with smaller litters.

Was the relatively lower lactational performance of mothers with large litters due to their inability to eat enough food, digest that food, and/or transport the energy and nutrients in that food across the intestinal wall to the blood? When food intake was compared to the summed uptake capacity of the small intestine, it was apparent that virgin females main-tained a reserve or excess uptake capacity that was more than 200 percent greater than their intake. At peak lactation, mothers still maintained an excess capacity for glucose uptake, but this time it was only 5–20 percent greater than glucose intake. Therefore, during normal lactation, the load on the gut, and in particular the load on the small intestine, was sub-stantially increased over that experienced by virgin mice. The capacity of the gut to utilize its load was not exhausted, however, even though the difference between load and capacity decreases with a higher energetic challenge.

Up to this point it was not known if mother mice had reached a ceiling on gut capacity, or whether they had been pushed to their maximal energy expenditure. Consequently, female mice were further challenged using acclimation to extreme cold temperatures (Konarzewski and Diamond 1994), and then acclimation to cold temperatures plus lactation (Hammond *et al.* 1994).

The coldest temperature that white mice can withstand, after acclimation for two weeks, is $-10\,°C$. As the acclimation temperature dropped from $23\,°C$ to $-10\,°C$, food intake and small intestinal glucose uptake capacity increased (Figure 8.2). The rate at which food intake increased, however, was faster than the rate at which glucose uptake capacity increased. Therefore, the excess capacity for glucose uptake declined to near zero at $-10\,°C$; however, the capacity of the gut was never exhausted. Interestingly, mice acclimated to $-10\,°C$ did not eat as much as lactating mothers with 14 pups at $23\,°C$ ($9\,g\,day^{-1}$ versus $20\,g\,day^{-1}$). Therefore, a cold energy challenge does not surpass the energy challenge of lactation in white mice in eliciting a maximal energy expenditure.

When lactating mothers were acclimated to $5\,°C$, the coldest temperature that mother mice with pups can withstand, they were able to eat an average of $3.9\,g\,day^{-1}$ more food than those lactating at $23\,°C$ (Figure 8.2), but they were not able to wean larger pups. Therefore, these mice had a greater energy intake, but were not able to increase their lactational performance. Finally, the small intestinal capacity of cold acclimated

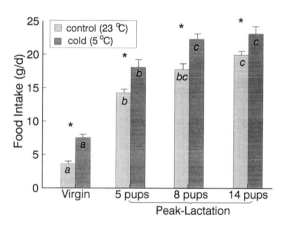

Figure 8.2.   Food intake versus reproductive demand of mice acclimated to either $23\,°C$ or $5\,°C$. Error bars indicate one standard error of the mean. Within a bar color, letters that are different correspond to means that are statistically different from each other. Asterisks indicate statistically significant differences between $23\,°C$ and $5\,°C$ mice within a level reproductive demand.

mice (both virgins and lactating females) was higher than that of control mice. Again, the excess capacity for glucose transport was near, but not equal to or below, zero. Thus, it is not the small intestine's capacity for glucose transport that limits lactational performance, and hence, whole-animal energy expenditure.

### Limitations on whole-animal energy expenditure during lactation

In the studies described above, a limit on the whole-animal energy expenditure was not observed, because the size of individual pups declines with increasing litter size. To test whether the limitation observed was related to milk production in the mammary glands, an experiment was performed involving the surgical removal of half the teats from an experimental group of lactating mice. Comparisons between control and experimental groups were made based on the pressure on each teat (pups/teat) during peak lactation. For example, experimental mice with 1.4 pups/teat had 7 pups and 5 teats, while control mice with 1.4 pups/teat had 14 pups and 10 teats; therefore, an experimental mother had the same load on each mammary gland, but less of a load on her gut compared to a control mother with the same pup to teat ratio. This manipulation was performed for teat pressures ranging from 1 pup/teat to 1.8 pups/teat. These experiments yielded two significant findings.

First, for a given mammary pressure, experimental mothers ate less food than control mothers. Thus, lactating females determine their daily food intake by the number of pups in their litter rather than by their maximal physiological capacity to assimilate food: lactating mothers are capable of eating more food, but they do not choose to do so.

Second, final litter masses increased nonlinearly, plateauing with increasing mammary pressure; but for a given mammary pressure the experimental mothers produce lighter litters than the control mothers (Figure 8.3). Hence, adding a pup to the litter results in a decreasing individual pup mass for all of the pups in the litter.

A logical site for this limitation, if not in the digestive tract or input processes, would be in the capacity of the mother to make enough high-quality milk to feed her pups. From studies in rats (Fiorotto *et al.* 1991), it appears as if the energy density of milk decreases in the largest litter sizes because milk is diluted with water in order to increase milk volume output. Therefore, the pups in the largest litters were smaller because they received a more energy-dilute diet. Because it is apparent that the limit to

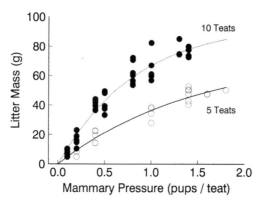

Figure 8.3. Litter mass versus mammary pressure for experimental (5 teats) or control (10 teats) lactating mice. The equation for each line follows the form litter mass $= a(1 - e^{-b})$ where $b$ is mammary pressure. The exponents for each line are as follows: 10 teats: $a = 96$, $b = 0.12$, $r^2 = 0.99$; 5 teats: $a = 73$, $b = 0.69$, $r^2 = 0.99$.

whole-animal energy output during lactation does not reside in the gut, it is probable that the limit on milk quality production, and thus output processes, explains the limit on whole-animal energy expenditure during lactation.

### Conclusions

These experiments demonstrate that the limit to whole-animal energy expenditure does not necessarily reside in the energy input pathways of digestion and intestinal transport capacity, but possibly in the energy output/milk production pathways. It has not been shown whether there is a limit on end-organ energy expenditure during maximal sustainable cold exposures or chronic exercise. It is clear that cold and lactation are additive in their effects on energy demand and intake, but these combined energy demands do not increase whole-animal energy expenditure beyond that of lactation alone (Hammond et al. 1994). Understanding if the limits to heat production and work are limiting to maximal metabolic performance during cold exposure or exercise is vital to understanding how various energy output pathways are matched to each other and to their respective energy demands, and to understanding if those pathways are optimized to meet the overall demands of the individual.

## Further reading

Fiorotto, M. L., Burrin, D. G., Perez, M. and Reeds, P. J. (1991) Intake and use of milk nutrients by rat pups suckled in small, medium, and large litters. *Am. J. Physiol.*, **260**, R1104–13.

Hammond, K. A. and Diamond, J. M. (1992) An experimental test for a ceiling on sustained metabolic rate in lactating mice. *Physiol. Zool.*, **65**, 952–77.

Hammond, K. A., Konarzewski, M., Torres, R. and Diamond, J. M. (1994) Metabolic ceilings under a combination of peak energy demands. *Physiol. Zool.*, **68**, 1479–506.

Hammond, K. A., Lloyd, K. C. K. and Diamond, J. M. (1996) Is mammary output capacity limiting to lactational performance. *J. Exp. Biol.*, **199**, 337–9.

Konarzewski, M. and Diamond, J. M. (1994) Peak sustained metabolic rate and its individual variation in cold-stressed mice. *Physiol. Zool.*, **67**, 1186–1212.

Oftedal, O. T. (1984) Milk composition, milk yield, and energy output at peak lactation: a comparative review. *Symp. Zool. Soc. Lond.*, **51**, 33–85.

Weiner, J. (1993) Physiological limits to sustainable energy budgets in birds and mammals: ecological implications. *Tree*, **7**, 384–8.

# 8.3 Optimization in design of the digestive system

IAN D. HUME

The tissues of the digestive system are expensive to maintain, and thus should be good candidates for optimization in design. However, optimization processes are often difficult to visualize, partly because of the operation of safety factors. The safety factor for any physiological process can be calculated as the ratio of its capacity to its maximal natural load. Measured safety factors fall in the range 1.2–10. Symmorphosis can be expected to be much more visible at the low end of this range than when a safety factor of 10 is operating; a large safety factor means large excess capacity most of the time.

Optimization processes may also be difficult to visualize when different anatomies evolve to perform the same or similar functions in different taxons, as can be illustrated in the evolution of digestive systems and digestive strategies in mammalian herbivores. The result is a diverse and sometimes confusing array of gut designs among different lineages that appear to utilize very similar plant food resources.

In the absence of barriers to the diffusion of enzymes through ingested food, the optimal design for a digestive tract is a tube (Penry and Jumars 1987). Its performance can be modeled as a plug-flow reactor (PFR), which features continuous input at one end, orderly flow with perfect radial mixing of reactants but negligible axial mixing within the reactor, and continuous output at the other end (Figure 8.4a). Consequently, there is a gradient of decreasing reactant concentrations but increasing product concentrations along the reactor. Plug-flow yields high rates of conversion of reactant to product per unit time, but the extent of reaction in practice is sensitive to barriers to diffusion such as lignified plant cell walls.

Consequently, nectarivorous birds, frugivorous bats, and insectivorous and carnivorous mammals, all of which feed on readily digestible

(a)

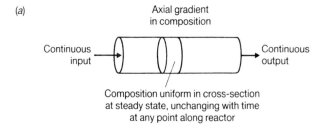

(b)

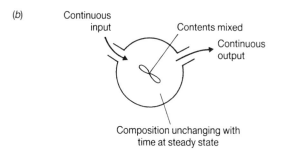

(c)

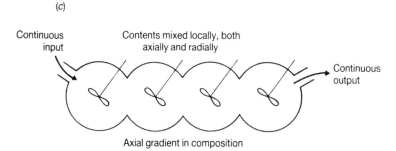

Figure 8.4. Diagrams of (a) an ideal plug-flow reactor; (b) a continuous-flow, stirred-tank reactor; and (c) a modified plug-flow reactor (a number of continuous-flow, stirred-tank reactors in series). (From Hume 1995.)

food sources, have digestive tracts dominated by a plug-flow reactor, the small intestine (Stevens and Hume 1995). The tracts of herbivores have one or more mixing chambers (continuous-flow, stirred-tank reactors (CSTR) in chemical engineering terms) in addition to a plug-flow reactor.

### The small intestine

In animals with plug-flow reactor guts, the rate of food intake can be limited by the rate of digestion. Physical limitations to rate of digestion are set by the capacity of the gut and its outflow rate, which set the upper limit to the volume of digesta that can be processed. Biochemical limitations are set by rates of enzymic hydrolysis and absorption in the small intestine. There is evidence for modulation of both types of limits in response to changes in energy demand and shifts in diet by the animal.

### *Modulation of gut capacity*

Physical limitations to digestion can be modulated by changing the length of the small intestine (and thus its volume and the transit time of digesta). For example, when house wrens (*Troglodytes aedon*) eating crickets were acclimated to either 24 °C or −9 °C, food intake at the lower temperature was double that at 24 °C, yet apparent digestive efficiency (metabolizability) of energy was significantly higher in the cold (77 percent versus 72 percent) (Dykstra and Karasov 1992). This was because the length and hence volume of the small intestine was 22 percent greater at −9 °C. The greater retention time in a longer small intestine, the absorptive region of the gut, permitted greater digestive efficiency without depressing food intake rate. The end result was a 150 percent greater rate of metabolizable energy intake to match the additional energy costs of thermoregulation at below-zero temperatures.

### *Modulation of enzyme activities and absorptive capacity*

Activities of the major digestive enzymes (amylase, proteases, and lipase) change in relation to the amounts of their respective substrates (starch, protein, and fat) in the diet. In rats, the activities of pancreatic amylase can increase five times, those of trypsinogen and chymotrypsinogen six times, and that of lipase eight times in response to the addition of, respectively, starch or glucose, protein, and fat or fatty acids, to the diet (Brannon 1990).

The small intestine's absorptive capacity is the product of mucosal surface area and nutrient uptake rates per unit area, summed over all regions of the small intestine. Principles of optimization suggest that activities of nutrient transporters in the brush border membrane of small intestinal villi should be modulated by diet composition. That is, a transporter should be repressed whenever its biosynthetic and maintenance costs exceed the benefits it provides. This prediction is only partly supported by experimental data, because of the operation of a number of other factors. The most important of these seems to be whether the substrate absorbed provides metabolizable energy or an essential nutrient. For instance, transporters for sugars tend to be upregulated by their substrates or by dietary carbohydrate. However, transport of water-soluble vitamins is often downregulated by increases in intake of the vitamin but upregulated in deficiency. Such a transporter is most needed at low intakes and least needed at high intakes when requirements might even be met by passive diffusion down a concentration gradient.

Modulation of both enzymic activity and absorptive capacity appears to be most rapid and effective in nectarivorous birds; all species of hummingbirds, honeyeaters, and lorikeets studied extracted at least 97 percent of the sucrose and glucose they ingested, even over a fourfold range of sugar concentration. Thus there appears to be a close match between load and capacity in the small intestine, and thus optimization in design. Furthermore, sucrase activity shows decreasing proximal to distal gradients along the small intestine matching the gradient in luminal sucrose concentrations as the sugar is hydrolyzed and absorbed. This is another example not only of optimization in a tissue which is expensive to maintain, but also of the characteristics of a plug-flow reactor, with decreasing concentrations of reactants along the length of the reactor.

## Fermentation chambers

The close matches between rates of digesta flow, enzymatic hydrolysis, and nutrient uptake found in the small intestine are not so readily apparent in the herbivore gut, which usually features one or more mixing chambers where microbial fermentation of plant cell walls takes place. Two types of fermentation chambers have been described for mammalian herbivores: continuous-flow, stirred-tank reactors (CSTRs) and modified plug-flow reactors (Hume 1989).

### *Continuous-flow, stirred-tank reactors*

Several mammalian herbivores have one or more mixing chambers proximal to their small intestine. These are foregut fermenters, and include ruminants, camelids, hippos, peccaries, kangaroos, wallabies and rat-kangaroos, sloths, and colobus and leaf monkeys. Mixing chambers are

also found distal to the small intestine, in the hindgut, of many mammalian herbivores.

Ideal CSTRs are spherical in shape, with separate ports for input and
output of materials. Both input and output are continuous, and the contents of the reactor are perfectly mixed (Figure 8.4b). Consequently composition is uniform throughout the reactor and unchanging with time at
steady state. The main disadvantage of a CSTR is that incoming reactants are instantaneously diluted by materials recirculating within the
reactor. Reaction rates are therefore lower than in a plug-flow reactor,
but the extent of reaction can be high if throughput rates are low enough
relative to volume (that is, if the retention time is long).

Clearly, the forestomach of foregut fermenters deviates from the ideal
CSTR in that feeding (input) is not continuous, and mixing is imperfect
because of internal elaborations of the wall of the mixing chamber.
However, outflow is modulated such that it is much more continuous
than food input, and performance approaches that of a CSTR (Hume
1989). Retention times of ingested food far exceed those in plug-flow
reactors, allowing time for substantial microbial degradation of plant
cell walls to occur.

The long retention times of ingesta in the ruminant forestomach mean
that when animals are given highly fibrous straw diets, the rate of clear-
ance of undigested residues is so low that the mixing chamber fills with a
mass of indigestible material and further input is inhibited (Figure 8.5).
Although digestibility is maximized, intake of digestible energy, or energy
absorbed, can fall below levels required for the animal's maintenance
(Hume 1989).

### Modified plug-flow reactors

Clearing the forestomach of undigested residues is less of a problem in
kangaroos, in which the forestomach functions as a number of CSTRs in
series. A CSTR can be improved by replacing it with two CSTRs in series
(with the same total volume), so that reactant concentrations do not fall
immediately to their final values. As the number of CSTRs in series
increases, the reactor system becomes more and more like a plug-flow
reactor (Penry and Jumars 1987), but still differs from a plug-flow reactor
principally because mixing is both radial and axial (Figure 8.4c). For
discussion purposes, tubiform fermentation chambers such as the kangaroo forestomach can be considered modified PFRs (Hume 1989).

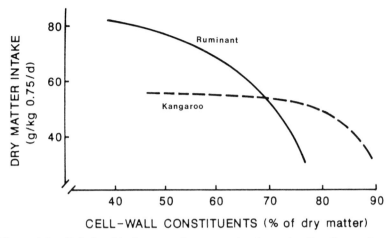

Figure 8.5. Relationship between dry matter intake and cell-wall constituents (fiber) content of forage for ruminants (solid line) and kangaroos (broken line). (Modified from Hume (1989) *Physiological Zoology*, **62**, 1145–63.)

That food intake constraints imposed by a high fiber content are less in kangaroos than in ruminants is shown by Figure 8.5. As fiber content increases above about 40 percent cell-wall constituents, food intake by ruminants decreases at an increasing rate. On lower fiber diets kangaroos eat 30 percent less than ruminants, reflecting their lower maintenance energy requirements, but on higher fiber diets they eat about 50 percent more than ruminants. While the ruminant strategy maximizes fiber digestibility, the kangaroo strategy appears to be to maximize intake of digestible energy at the expense of fiber digestibility; fiber digestibility is consistently less in kangaroos than in sheep fed the same diet (Hume 1989).

Another fermentation chamber that functions as a modified PFR (that is, a number of CSTRs in series) is the equine proximal colon. Both the proximal colon of the horse and the forestomach of the kangaroo are tubiform in gross morphology (Figure 8.6). Like kangaroos, horses are better able to maintain food intake as dietary fiber content increases than are ruminants (Hume 1989).

### Evolution of herbivores and optimal gut design

Several examples of convergent evolution between marsupial and eutherian herbivores suggest that optimization of gut design is based on some

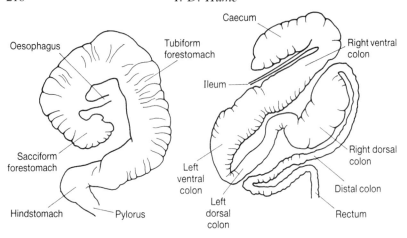

Figure 8.6.  The kangaroo forestomach (left) and the equine hindgut (right) as examples of modified plug-flow reactors. (From Hume 1995.)

common principles. The modified PFR fermentation chambers in the kangaroo forestomach and the equine proximal colon provide one such example: both appear to have evolved to their present form, albeit at different ends of the gut, in response to the spread of grasslands as world climates dried and cooled in the Miocene epoch, and the problem of maintaining food intake as grasses became more fibrous.

A second example is the solution to the fecal loss of essential amino acids and B-vitamins synthesized by the microbes in the caecum of small hindgut fermenters (caecum fermenters). Both the marsupial ringtail possums (*Pseudocheirus* spp.) and eutherian rabbits ingest high-nutrient feces derived from caecal contents, thereby salvaging valuable nutrients that would otherwise be lost. The evolution of caecotrophy by both groups of herbivores has enabled them to exploit much more fibrous diets than their small body size would suggest was possible.

## Conclusion

The concept of optimization in the design of the digestive system is supported by close matches observed between rates of digesta flow, enzymatic digestion, and nutrient absorption in the small intestine. Optimization is more difficult to demonstrate in the range of fermentation chambers in mammalian herbivores. However, analysis of the function of these fermentation chambers in relation to optimization processes is facilitated by reference to the performance of chemical reactors, especially continuous-flow, stirred-tank reactors (CSTRs) and modified plug-flow reactors, which are a number of CSTRs in series.

### Further reading

Brannon, P. M. (1990) Adaptation of the exocrine pancreas to diet. *Annual Review of Nutrition*, **10**, 85–105.

Dykstra, C. R. and Karasov, W. H. (1992) Changes in gut structure and function in house wrens (*Troglodytes aedon*) in response to increased energy demands. *Physiological Zoology*, **65**, 422–42.

Hume, I. D. (1989) Optimal digestive strategies in mammalian herbivores. *Physiological Zoology*, **62**, 1145–63.

Hume, I. D. (1995) Flow dynamics of digesta and colonic fermentation. In *Physiological and Clinical Aspects of Short-Chain Fatty Acids*, pp. 119–32. Eds. J. H. Cummings, J. L. Rombeau and T. Sakata. Cambridge University Press, Cambridge.

Penry, D. L. and Jumars, P. A. (1987) Modeling animal guts as chemical reactors. *The American Naturalist*, **129**, 69–96.

Stevens, C. E. and Hume, I. D. (1995) *Comparative Physiology of the Vertebrate Digestive System*, 2nd edn. Cambridge University Press, Cambridge.

## 8.4 How ruminants adapt and optimize their digestive system "blueprint" in response to resource shifts

### REINHOLD R. HOFMANN

Mammalian herbivores face a variety of problems in tapping their main sources of nutrition and in meeting their energy costs from the plants they digest. Terrestrial plants offer cell solubles which are completely available and cell wall components composed of several structural carbohydrates, such as cellulose, which can be broken only indirectly by providing large intestinal fermentation vats adopted by specific populations of anaerobic microbes releasing cellulase and thus producing short-chain fatty acids that are absorbed by the forestomach or hindgut mucosa.

Ruminants, as "late-comers" in mammalian evolution have – over the past 25 million years or so – approached their forage selection and plant cell wall digestion problems with a series of interdependent morphophysiological devices. Their main fermentation chamber is proximal in the foregut, in contrast to most other eutherian herbivores where it is in the hindgut. Plant evolution, especially the radiation of grasses during the miocene period less than 14 million years ago, has clearly influenced ruminant evolution as cellulose then became the most abundant carbohydrate. The wide-ranging availability of this immense biomass of grasses has, since then, promoted an almost explosive radiation, mainly of bovids.

### Gradual shift from plant cell content to plant cell wall

During this long-lasting evolutionary period, gradually changing from plant cell content selection to general plant cell wall utilization, several optimized digestive structures were reduced or given up because they became functionally redundant. Others, originally poorly developed in the "blueprint" of early (usually small) ruminants, sprang up.

Ruminants ferment plant cell wall and storage forms of plant cell contents like starch in a highly differentiated non-glandular forestomach system with an absorptive cutaneous mucosal lining. It is served by well-developed salivary glands (without an enzymatic function) opening in the mouth cavity, which have a high carbonate and phosphate buffering capacity since SCFA releasing rumen bacteria function optimally near neutral pH. Ruminants chew the cud, that is they ruminate in order to break down especially fibrous forage constituents into very small particles (0.5–2.0 mm), optimizing the access of microbes and their cellulolytic enzymes to the plant structural carbohydrates.

The digestion of certain cell wall structures (leaf hemicelluloses) requires maceration by HCl, which is produced only in the fourth portion of ruminants' quadrilocular stomach, the abomasum. It corresponds to any mammalian glandular stomach, by secreting pepsinogen in its chief cells and hydrochloric acid in its parietal cells. Considerable proportions of the abomasal HCl are used as a bactericide for subsequent proteolysis of the "self-produced" rumino-bacterial protein. HCl is also required, apart from use in maceration of structural carbohydrates as mentioned, for the neutralization of alkaline saliva flowing into the abomasum directly from the mouth via a sophisticated shortcut, bypassing the entire forestomach fermentation system. Its initial function in the suckling ruminant is to direct milk into the glandular stomach and not into the (proportionally still small) reticulorumen. This is achieved in the suckling ruminant via a reflex closure of the lips of this ventricular groove and of the openings towards the forestomach cavities. This reflex is lost after weaning and, for a long time, the groove was thought to be non-functional in the adult.

Due to its high structural development, allometric growth and great functional potential in all adult ruminants, I hypothesized more than 25 years ago (Hofmann 1968a) that this bypass (sulcus ventriculi) is used essentially by all adult ruminants to transport soluble nutrients (plant cell contents, PCC) released during initial mastication (and diluted by excessive serous saliva) into the abomasum. These readily digestible solubles, especially glucose, would otherwise be degraded (and "wasted") by the rumen bacteria. During initial mastication of fresh juicy plant material, and presumably helped by the saliva-releasing effects of plant secondary compounds like tannins, however, substantial amounts of soluble plant glycerols, proteins, and sugars can be washed into the abomasum. Wild ruminants selecting their forage primarily for such valuable PCCs have retained expression of intestinal sodium-glucose co-transporter (Rowell *et al.* 1996). The smaller they are, the less they can gain their energy requirements from rumen fermentation only (Van Soest 1994); they rely on bypass.

This provides ruminants with a third avenue to digest nutrients, which is comparable to that of animals with a simple stomach. It functions in addition to forestomach and hindgut (that is, fractionated) fermentation. The malleability of the "new" forestomach system in relation to evolutionary shifts in forage resource distribution and diversification of ecological niches is astounding and reduces restraints to extreme seasonal or climatic conditions only.

Ruminants are extremely diverse in shape and body mass (4 to 1,200 kg), and they can adapt to a multitude of extreme environments – in both tropical and temperate or even arctic climates. Their basic morphophysiological principles were obviously established millions of years ago, but their variation steps as demonstrated by *c.* 180 taxonomically diverse extant species, resemble a baobab tree: on every level of specialization and adaptation to their specific forage, ruminants are all highly developed herbivores.

During my extensive studies of the rich East African ruminant fauna between 1962 and 1972 (Hofmann 1968b, 1973) I observed marked morphological differences in the stomachs and salivary glands of small, medium-sized and large species, from gross anatomy down to ultrastructural level. This was related to botanical studies and feeding behavior observations. The division into a flexible system of three partly overlapping feeding types – concentrate selectors (CS), intermediate mixed feeders (IM), and grass and roughage eaters (GR) (Hofmann and Stewart 1972) – provided a morphophysiological framework for appreciating ruminant diversification, specialization and, in my opinion, optimization of the structure–function relationship and its shifting emphasis along the tract in relation to the plant resource a ruminant is utilizing (Figure 8.7). We later verified and extended these findings in European, Asian, and American species and provided morphological data for all portions of the digestive system.

Ruminants of the three feeding types occupy different nutritional niches although they often occur sympatric and practice complementary resource utilization (for example, in Africa). They face different principal and seasonal problems and therefore also show different morphophysiological responses in their varying optimization strategies. These

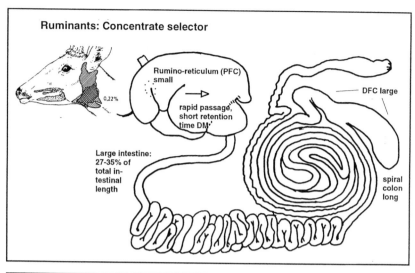

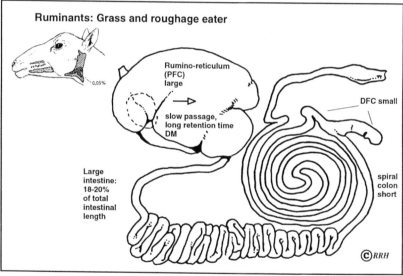

Figure 8.7.  Semischematic representation of physiologically important portions of the ruminant digestive system in a small (30–40 kg) CS cervid (roe deer, above) and a small (30–40 kg) GR bovid (mouflon, below); 0.22% and 0.05% refer to parotid glandular mass as a proportion of body mass. DM, dry matter (natural forage); DFC, distal fermentation chamber (caecum and proximal colon loop); PFC, proximal fermentation chamber (rumino-reticulum).

differences are forced upon them by plant structure, especially variable cell wall digestibility or plant cell contents availability, or suitable forage availability in general (for snow, overgrazing, monoculture, and so on), and, last but not least, by plant secondary (defense) compounds (Owen-Smith 1982).

### Adaptation in grazers (GR)

Our image of ruminants has been biased by countless investigation results obtained from sheep and cattle, both with firm evolutionary adaptations to seasonally lignifying temperate grasses as the main forage resource, with cattle (domesticated from the extinct Eurasian aurochs) showing the most advanced grazer (GR) features: nonselective forage prehension (bulky tongue against rigid upper lip), heavy mandible attaching widespread, overcrossing masticatory muscle portions optimized for sideways grinding with accordingly adapted broad selenodont cheek teeth (complete breakdown of fibrous particles); and salivary glands, producing in cattle up to 180 liters of buffering and diluting rumen fluid per day. It contains, however, no specific protein that could neutralize one particularly diverse group of plant secondary compounds, that is, the tannins. Their glands have a much higher proportion of mucous salivary cells than in CS, and they show a higher degree of oral mucosa cornification – all adaptations to potentially hard, fibrous forage. Total salivary gland mass is about half of a CS of similar weight.

Full development of a voluminous reticulorumen as the main (proximal) fermentation chamber occupies 65 percent of the abdominal cavity. Subdivision by strong, muscular pillars and narrow ostia provides sufficient delay for cellulolysis (which depends on a long retention time!). All this results in low fermentation rates, a relatively small rumen mucosal surface enlargement by papillae mainly at midlevel, and reduced blood flow in uppermost and lowermost rumen areas (coarse forage stratification).

In free-ranging ruminants, changes in forage quality (for example, increased digestibility in the spring or rainy season, and reduced digestibility in the winter or dry season, due to lignification) cause microbial population shifts which influence SCFA composition and fermentation rates: within an adaptation period of two to three weeks, rumen papillae decrease or increase in length and number, and parotid glandular tissue proliferates or atrophies.

Since most GR are adapted to low fermentation rates with slow though steady absorption of SCFA and gradual gluconeogenesis activities of the hepatocytes, their liver amounts to only about 1.1–1.3 percent of body mass. The intestine is usually very long (20–30 times body length), with the small intestine amounting to 75–82 percent of total intestinal length, and the large intestine to only 18–25 percent. The typical spiral colon of a ruminant is relatively short (unless used excessively for water reabsorption as in the oryx) and the widened caecum with proximal colon loop is poorly developed as a "distal fermentation chamber" (DFC) with a ratio of 1 : 15–30 in relation to the reticulorumen, the proximal fermentation chamber (PFC).

### Concentrate selectors (CS) still face the problem of fibrous forage

This early evolved group (40 percent of all ruminant species) comprises the small fruit and dicot selectors (for example, dikdik, duiker, and roe deer), and the larger bush and tree foliage selectors or browsers (for example, moose, kudu, and giraffe). Evolved before GR, their continued survival on various continents today reflects the prevailing habitat types: bush versus grass savanna in Africa, or forest versus steppes or prairies in Eurasia or America. Many of them are territorial.

When CS evolved and adapted, well before grass radiation, plant secondary compounds (PSCs) must have been their main nutritional constraint. At the same time, though, PSCs made up part of the plant cell contents; that is, what ruminants of this feeding type specifically select.

Since they select against plant cell wall (fiber), yet often their selected diet automatically includes high PSC levels, these browsing species must have well-developed cellular mechanisms for dealing with these PSCs: high proportions of specific binding proteins in their saliva neutralize, for example, condensed tannins; and we have seen how differentiated they exploit cell contents in general. The smaller a CS species, the larger (in relation to body mass) are its serous salivary glands: up to 0.25 percent of body mass.

CS have difficulty in selecting suitable juicy forage in winter or during the tropical dry seasons, because it is simply not available. They compensate by selecting PSC-rich plants rather than embarking on a greater intake of lignified/fiber-rich forage which they cannot digest. Whilst they do not reduce particle size as efficiently as GR, as long as they can afford to be selective, they ruminate very thoroughly in winter when they are large like moose (Nygren and Hofmann, 1990) and practice normal cellulolysis in their relatively small rumen, although on a small scale – obviously complementing these small energy gains with fat reserve catabolism, ventricular groove bypass, and increased hindgut fermentation/ SCFA production. In these periods of reduced forage quality and intake, their salivary glands atrophy; in moose, for example, the glands atrophy to 60 percent of the summer state (Hofmann and Nygren 1992).

CS have a relatively smaller, simpler structured reticulorumen (empty weight 1.5 percent of bm versus 2.5–3.5 percent in GR) with weak pillars and wide openings, leading to reduced retention time. There is no stratification of the well-squashed forb or foliage ingesta which are increasingly diluted by copious saliva flow. This additionally reduces MRT (mean retention time) in the rumen and thus induces a higher forage intake frequency. High fermentation rates, as long as selectivity for high-quality forage can be practiced, maintain in CS an even rumen papillation with a generally high surface enlargement factor of at least ten times, frequently above twenty times.

The omasum, the final differentiation product of mammalian stomach evolution found only in ruminantia, is small and simply structured in CS but large and highly differentiated in GR. The small CS omasum offers, through a few small laminae, only little surface enlargement. In a 280 kg moose the surface is only 1,820 cm$^2$, compared to 40,000 cm$^2$ in a 600 kg cow as an example of GR.

All CS have a relatively big liver (1.9–2.6 percent of bm, which is twice as much as in GR) with little connective tissue. Not surprisingly, liver mass shows metabolic/seasonal changes in the CS feeding type more than in the other two. Its relevance must be higher to PSC than to fiber digestion.

The CS intestine, finally, is shorter than in GR, but its hindgut portion is at 27–35 percent relatively longer and much more adapted to secondary fermentation of, for example, hemicellulose and absorption of nutrients, water, and electrolytes/minerals. The spiral colon is proportionally longer; it completes the typically diverse function of the tract, in this case efficient water-reabsorption.

## Mixed feeders are successful digestive opportunists

Mixed feeders (IM) (35 percent of extant species) show many intermediate morphological signs, adaptive emphasis shifting seasonally from a more CS-typical strategy with corresponding morphological changes to a more GR-oriented adaptive swing, or vice versa (Figure 8.8). For

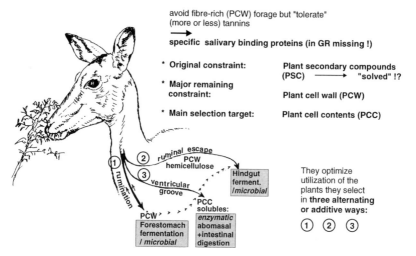

**Wild, free-ranging RUMINANTS: selective IM and CS**

avoid fibre-rich (PCW) forage but "tolerate"
(more or less) tannins

→

specific salivary binding proteins (in GR missing !)

* Original constraint:          Plant secondary compounds
                                (PSC) ——→ "solved" !?

* Major remaining
  constraint:                   Plant cell wall (PCW)

* Main selection target:        Plant cell contents (PCC)

ruminal escape
PCW
hemicellulose

Hindgut
ferment.
/microbial

Ventricular
groove

PCC
solubles:

PCW

enzymatic
abomasal
+intestinal
digestion

Forestomach
fermentation
/ microbial

They optimize
utilization of the
plants they select
in **three alternating**
or additive ways:

① ② ③

Figure 8.8.   Most nondomestic ruminants are selecting forage according to its quality mainly by olfaction unless they are grazers. Depicted is an opportunistic intermediate ruminant – an impala antelope browsing from an acacia bush. Abbreviations: CS, concentrate selector; GR, grass and roughage feeder (grazer); IM, intermediate mixed feeder; PCC, plant cell contents (soluble nutrients – sugars, lipids, and so on – but also tannins); PCW, plant cell wall constituents (cellulose, hemicellulose, pectin; that is, "plant fiber"); PSC, plant secondary compounds (polyphenols, for example, condensed tannins, diterpenes, and so on).

example, Impala antelopes, like most cervinae species, give up their growing-season preference for grass species admixed with forbs, when lignification advances, in favor of PSC-loaded acacia foliage in the dry season, or tree foliage buds, litter, acorns, and conifer needles in winter respectively. All species are selective and avoid fiber as long as possible. They change digestive avenues to alleviate seasonal restraints of forage quality and availability. They have developed strong social systems which are clearly related to their successful feeding strategies.

## Conclusions

The blueprint of the ruminant digestive system is expressed in 180 extant species through its adaptive variations. They are matched to different functional requirements in contrasting ecological niches. Evolution of the blueprint structures has resulted in contrasting feeding types including transitional (intermediate) forms. A concentrate selector cannot become a grazer or vice versa – in spite of man's forceful attempts, for example cattle fed with cereals and animal proteins.

Optimization of "typical ruminant" structures has changed the prehensile, masticatory, and salivary apparatus as much as the gastrointestinal tract, that is, soft tissues (Hofmann 1988, 1989) and skeletal structures (Spencer 1995).

For most ruminant species, insufficient morphophysiological adaptation to high fiber forage or lignified PCW is the much stronger constraint than for CS and IM to overcome the plant defense systems, both chemical or mechanical. We are far from knowing exactly how ruminants of the various feeding types cope with a frequently hostile and changing environment – many biochemical and physiological data complementing our morphological observations are still missing and are difficult to obtain from wild ruminants. Feeding them under controlled conditions with standardized rations means upsetting their complex biocontrol system with its unpredictable interdependencies, and misinterpreting their considerable adaptive tolerance to feeds they never would select on free range, in their nutritional niche.

It is in this context that body size is used by biostatisticians to explain or even out differences between various ruminant species. Since, however, CS and IM continue, on free range, to select plants loaded with PSC instead of adapting to abundant grass forage while GR on overgrazed range will starve to death in reach of PSC loaded browse to which they are not adapted, body size cannot fully explain the variation of ruminants. Their further evolution, however, is more and more in question. This also means the potential destruction of a multifaceted, prime example for adaptive structural changes to fit altered functional demands, the essence of the concept of symmorphosis.

### Further reading

Hofmann, R. R. (1968a) Zur Topographie und Morphologie des Wiederkäuermagens im Hinblick auf seine Funktion (nach vergleichenden Untersuchungen an Material ostafrikanischer Wildarten). *Zentralblatt Vet. Med.*, Beiheft Nr. 10, 1–180. Paul Parey Verlag, Berlin and Hamburg.

Hofmann, R. R. (1968b) Comparisons of rumen and omasum structure in East African game ruminants in relation to their feeding habits. Symp. Zool. Soc. Lond., No. 21, *Comparative Nutrition of Wild Animals*, pp. 179–94. Ed. M. A. Crawford.

Hofmann, R. R. (1973) *The Ruminant Stomach, Stomach Structure and Feeding Habits of East African Game Ruminants*. East African Monographs in Biology, vol. 2, pp. 1–364. E.A. Lit. Bureau, Nairobi.

Hofmann, R. R. (1988) Morphophysiological evolutionary adaptations of the ruminant digestive system. In *Aspects of Digestive Physiology in Ruminants*, pp. 1–20. Eds. A. Dobson and M. Dobson. Cornell University Press, Ithaca and London.

Hofmann, R. R. (1989) Evolutionary steps of ecophysiological adaptation and diversification of ruminants: a comparative view of their digestive system. *Oecologia*, **78**, 443–57.

Hofmann, R. R. and Nygren, K. (1992) Morphological specialisation and adaptation of the moose digestive system (*Alces alces* L.). Proc. 3rd Int. Moose Symp., *ALCES*, Suppl. 1, 77–83.

Hofmann, R. R. and Stewart, D. R. M. (1972) Grazer or browser: a classification based on the stomach structure and feeding habits of East African ruminants. *Mammalia, Paris*, **36**(2), 226–40.

Lechner-Doll, M., Hume, I. D. and Hofmann, R. R. (1995) Comparison of herbivore forage selection and digestion. In *Recent Developments in the Nutrition of Herbivores*, pp. 231–48. Eds. M. Journet, E. Grenet, M.-H. Farce, M. Theriez and C. Demarqulilly. Institut National de la Recherche Agronomique, Paris.

Nygren, K. and Hofmann, R. R. (1990) Seasonal variations of food particle size in moose. *ALCES*, **26**, 44–50.

Owen-Smith, N. (1982) Factors influencing the consumption of plant products by large herbivores. In *The Ecology of Tropical Savannas*, pp. 359–404. Eds. B. J. Huntley and B. H. Walker. Springer Verlag, Berlin, Heidelberg and New York.

Owen-Smith, N. and Cumming, D. H. M. (1993) Comparative foraging strategies of grazing ungulates in African savanna grasslands. In *Grasslands for our World*, vol. 1, pp. 691–8. Ed. M. J. Baker. SIR Publishing, Wellington, NZ.

Rowell, A., Meyer, H. H. D., Shirazi-Beechey, S. P. and Hofmann, R. R. (1996) *Abstract Vol., 1st Int. Symp. Physiol. Ethol. Wild and Zoo Animals*, p. 91. Schriftenreihe Institut Zoo- und Wildtierforschung, Berlin, ISSN 1431–7338.

Spencer, L. M. (1995) Morphological correlates of dietary resource partitioning in African Bovidae. *Journal of Mammalogy*, **76**, 448–71.

Van Soest, P. J. (1994) *Nutritional Ecology of the Ruminant*, 2nd edn. Cornell University Press, Ithaca and London.

## 8.5 Optimality in complex dynamic systems: constraints, trade-offs, priorities

### WOLFGANG WIESER

I think most of us would agree that principles of economy and optimality have played a decisive role in the evolution of biological systems, and that the concept of symmorphosis, insofar as it refers to such principles, is a plausible concept. How else but by the economy and optimality of design could one explain the extraordinary efficiency of biological systems – the fact, for example, that the basal power requirements of a human being are of the order of 100 watts, equivalent to that of a light bulb? Still, it has been found difficult to define optimality as a general scientific principle because in biological systems it cannot be falsified. This has to do with the complexity, interconnectedness, and dynamic nature of biological systems which make it impossible to decide whether failure to predict the shape of the relationship between a particular function and a particular structure, or between different functions, signals the falsification of a hypothesis or is simply due to the model builder having overlooked important constraints, or being unable to provide a stringent definition of the boundary conditions of the system under study.

The serendipitous discovery of a hidden constraint may require the introduction of a new parameter into an old model, thus altering the framework within which predictions on the optimality of design can be made; and there is no way of knowing when the end of the line has been reached.

In what follows I shall present two examples of how our confidence in predictions about energy acquisition and energy allocation in cells and organisms has been affected by new data and by the discovery of hitherto unrecognized constraints.

## Switching strategies: ontogeny of temperature relationships in fish

In aquatic organisms an increase in water temperature often signals the advent of favorable environmental conditions in spring, promising a plethora of food. For the organisms concerned it is advantageous to divert as large a proportion of the seasonal bloom as possible into body substance, thus accelerating either growth or reproductive processes. One way of exploiting the possibilities offered by high temperature and thus the promise of abundance is to let the rate of energy input increase more steeply with temperature than the rate of energy expenditure, thus increasing the amount of food energy potentially available for production, that is, to increase the "scope for growth" (Brett 1976).

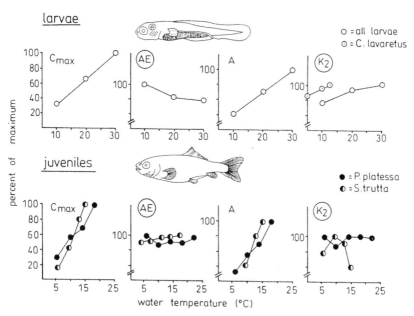

Figure 8.9. Effects of water temperature on components of energy budgets of two ontogenetic stages of fish, expressed in relative terms (maximum value = 100 percent). All values are based on energy equivalents. $C_{max}$ = maximum rate of ingestion; $A$ = maximum rate of assimilation = $P_{som} + R_{tot}$, where $P_{som}$ is the maximum rate of growth and $R_{tot}$ is the total metabolic expenditure; $AE$ = assimilation efficiency; $K_2$ = net conversion efficiency; that is, $P_{som}/A*100$. Data are based on the following references: all larvae: Houde (1989); *C. lavaretus* (whitefish): Hanel *et al.* (1996); *P. platessa* (plaice): Fonds *et al.* (1992); *S. trutta* (brook trout): Elliott (1976).

The lower part of Figure 8.9 shows how in the juveniles of two of the best studied species of fish, brook trout (Elliott 1976) and plaice (Fonds *et al.* 1992), flux and allocation of food energy respond to an increase in temperature. Within the preferred range the temperature coefficients of maximum food uptake ($C_{max}$) and rate of food assimilation ($A$) are uniformly high, whereas the two quotients of energy allocation, i.e. assimilation efficiency ($AE$) and net conversion efficiency ($K_2$, according to Ivlev) – are more or less unaffected by temperature. One may be justified in speculating that in these animals energy input into production is determined entirely by factors determining *flux* rather than by factors altering the pattern of *allocation* (see Wieser 1994), suggesting that the rate of maximum growth is ultimately controlled by the capacity for food uptake. Fish larvae, after beginning to feed on zooplankton, grow much faster than juveniles (Wieser 1995), although at this time their digestive apparatus is not yet fully developed. The lack of a stomach and the fact that the gut is short and straight are responsible for fast evacuation rates and a general tendency to lose digestive enzymes with the feces. The question arises as to how this particular set of constraints influences the choice of the best strategy for maximizing growth performance.

Comparing the two ontogenetic stages of fish illustrated in Figure 8.9, it emerges that the temperature coefficients of the two flux components, $C_{max}$ and $A$, were lower in the larvae than in the juveniles, and that, contrary to the situation in fully developed fish, assimilation efficiency decreased and net conversion efficiency increased with temperature in the larvae. It appears then that in larvae, under favorable thermal conditions, energy input into production is determined not only by the temperature-dependent increase in flux but also by a change in the way assimilated energy is allocated. This strategy could not have been predicted from a model based exclusively on experiments with juveniles.

However, Figure 8.9 implies that by this switch in strategy not enough power may be gained to account for the huge increase in the rate of growth in larvae as compared to juveniles. Whereas, for example, at the preferred temperature relative growth rates in juvenile brook trout weighing a few grams are of the order of a few percent, the larvae of roach (*Rutilus rutilus*) and whitefish (*Coregonus lavaretus*) weighing between 1 and 10 mg may grow at 25–30 per cent per day. Attempts to fit growth rates of this magnitude into conventional energy budgets uncovered yet another constraint and presented us with a new difficulty.

## Energy allocation at maximum performance levels: additions and trade-offs

Fitting the metabolic cost of growth into an animal's energy budget relies on a few basic principles of which the most relevant for our present purpose are the following:

(i) the metabolic cost of growth (here abbreviated $R_g$) is added to the cost of maintenance ($R_m$) of the organism;

(ii) provided the composition of the body remains constant, $R_g$ is directly proportional to the rate of growth ($g$);

(iii) the relationship between mass-specific $g$ and body mass ($G$) is defined by a power law of the form $g = a^* G^{-b}$, so that $\log g = \log a - b^* \log G$.

In Figure 8.10 these basic principles are illustrated by experimental data on fish. The upper panel (1) embodies principles (i) and (ii), based on experiments with juvenile fish growing at less than 10 percent per day. The middle panel (2) represents principle (iii), the upper curve being based on experiments with the larvae of *Rutilus rutilus* and *Coregonus lavaretus* on maximum ration, their growth rates always exceeding 13 percent per day. The lower curve represents a hypothetical $R_g$ calculated by means of the calibration curve shown in panel (1). From the functional relationships assembled in the lower panel (3), it becomes apparent that in the larvae the experimentally determined values of $R_g$ diverged widely from those calculated on the strength of the growth performance of juveniles, thus leading to a considerable difference between measured and calculated $R_{max}$ values. The discrepancy between measured and calculated costs is best illustrated by separating the metabolic expenditure for maintenance from that for growth (Figure 8.11). Whereas the lower panel of Figure 8.11 indicates that the measured mass-specific rate of maintenance metabolism in the larvae decreased with the conventional exponent of about $-0.2$ (Wieser 1995), the upper panel seems to suggest that the measured cost of growth was independent of body mass and thus of growth rate in larvae ranging in weight from about 2 to about 300 mg (wbm). The hatched area in the upper panel illustrates the difference between the measured values of $R_g$ and those expected on the basis of the measured rates of larval growth and the calibration curve for juveniles (Figure 8.10, panel 1).

Since protein concentration, the most costly component of somatic growth, remained constant in the larvae measured, we have to look for

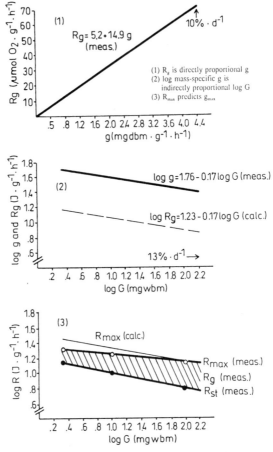

Figure 8.10. Relationships required for calculating the mass-specific cost of growth in fish of different size: (1) cost of growth ($R_g$) plotted against rate of growth (g) in juvenile fish growing at < 10 percent per day (based on Wieser 1994); (2) logarithmic transformation of the mass relationship of the measured rate of growth and of the calculated corresponding metabolic cost in larvae of *R. rutilus* (roach) and *C. lavaretus* (whitefish) growing in excess of 13 percent per day; (3) logarithmic transformation of the mass relationship of measured and calculated components of metabolic expenditure of larvae of *R. rutilus* and *C. lavaretus* at 20 °C. $R_{st}$ = standard metabolic rate; $R_{max}$ = maximum rate; $R_g = R_{max} - R_{st}$ = cost of growth (hatched area); G = body mass.

a source of power for the apparent deficit emerging from a comparison between measured and expected costs. It has been suggested (Wieser 1994, 1995) that a temporary reduction of maintenance functions (particularly of protein turnover) may create a power surplus that could be traded off against the power missing for the fueling of fast

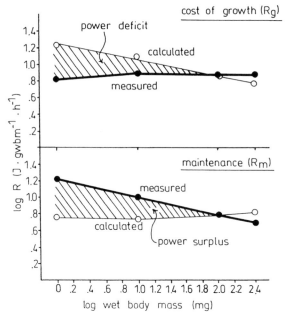

Figure 8.11. As for Figure 8.10 (panel 2), but separate, noncumulative representation of the mass relationship of the rate of maintenance metabolism and of the cost of growth. "Power deficit" and matching "power surplus" represent the difference between measured and calculated metabolic expenditure over a range of measured rates of growth, corresponding to a range of body weights as shown in Figure 8.10 (panel 2).

growth. This trade-off is illustrated by the two hatched areas in Figure 8.11. In the smallest larvae investigated, weighing about 2 mg (wbm), the trade-off would involve about 35 percent of the maintenance expenditures of non-growing larvae. With the increase in body mass and the concomitant slowing of growth performance the power deficit decreased, disappearing at a larval body weight of about 100 mg.

It is obvious that the maximum growth performance of fish larvae could not have been derived from the conventional additive model of energy allocation, any more than the efficiency with which food energy was converted into body mass could have been predicted from measured growth rates alone.

## Discussion

The message of case no. 1 presented here is quite simple: different rules of optimality, economy, and efficiency may apply to different stages of life cycles. However, the case of the developing fish offers a few interesting and, as it transpires, unexpected features. Fish larvae at hatching are the smallest free-living and free-feeding vertebrates and may increase in body mass by three orders of magnitude within one month or so (Wieser 1995). Having to acquire a protein-rich diet as rapidly as possible demands a carnivorous lifestyle, which in turn requires that the larvae be equipped with reasonably efficient organ systems for locomotion, prey detection, prey capture, digestion, and respiration. In contrast to holometabolous insects, the larvae and adults of fish represent the same life form so that the tissues and organs of the larvae form the morphophysiological basis on which the adult form must be built.

This set of premises contains the roots of ontogenetic conflicts between growth and differentiation: on the one hand, there is selective pressure for fast growth which means, in essence, fast deposition of protein in the inner muscle mass; on the other hand, in order to achieve this goal differentiation of the organ systems necessary for successful hunting and food acquisition also has to proceed sufficiently fast. Compromises are to be expected, like delaying the development of some organ systems. Indeed, the delayed differentiation of the gut and the digestive system may be a major cause for the differences in temperature relationships of energy acquisition and allocation between larvae and juveniles illustrated in Figure 8.9.

For the purpose of a more general discussion the assumption is made that there is an ontogenetic timetable by means of which the sequential pattern of differentiation of organ systems is adjusted to the changing needs and capabilities of the growing fish. To complicate matters, this timetable also depends on cues from the environment, transmitted mainly via food and temperature (Johnston 1993).

Case no. 2 again seems to illustrate an almost trivial principle, namely that all the ATP-consuming functions of an organism cannot proceed at maximum rates at the same time. It is self-evident that, for example, a fish swimming at the maximum velocity cannot also grow at maximum rate. In fact, in such a fish more than 90 percent of all the oxygen taken up is shunted to the swimming muscles so that other physiological functions, like digestion and synthesis, are impaired (Randall and Daxboeck 1982). However, the trade-off illustrated in Figures 8.10 and 8.11 is of a more fundamental nature. It implies that not even the organismic and cellular functions collectively pigeonholed under the term "maintenance," can be considered invariable, fixed components of energy budgets, but may slow down or even stop in the face of an additional and dominant demand. In this way the metabolic demands of maintenance functions are believed to be traded off against the metabolic demands of fast growth in fish larvae (Wieser 1994). This is in marked contrast to the behavior of, for example, the mice studied by Hammond *et al.* (1995), in which even the combination of several peak energy demands did not exhaust the capacity of a strictly additive "scope for metabolic performance."

On the basis of the findings with fish larvae I am tempted to favor, at least for small and active poikilothermic animals, a discontinuous, non-additive model of energy allocation in which metabolic power is not partitioned to different functions according to the free capacity of a fixed scope for metabolic performance, but according to shifting priorities of demand. This immediately raises the question of the temporal structure of such a dynamic model. For how long, for example, could the maintenance function believed to be traded off against the metabolic cost of growth in fish larvae, remain suppressed?

The common theme of the two examples discussed here is that in complex dynamic systems the allocation of energy to the functions consuming free energy must also depend on programs for the optimal *allocation of time* to each function. This may involve the existence of timetables determining the sequence of differentiation and growth of different organs during ontogeny (case no. 1), or dynamic programs monitoring and controlling priorities of demand in the organism and its cells, on which the temporal structure of energy allocation is based (case no. 2). Such dynamic programming may result in shifting patterns of energy flow, leading to discontinuities in the rates of some of the functions involved. The finding that in young vertebrates growth proceeds in a discontinuous, aperiodic saltatory fashion may be a consequence of this principle.

It is interesting that even though all biologists are aware of the fact that organisms are open dynamic systems, models of functional relationships in organisms often stress (or imply) the *stability* rather than the *flexibility* of structures, networks, set points, optimal solutions, final states, and so on. It should be remembered, however, that the constancy of parameter values by which a system in steady state is defined is achieved only by the variance and dynamic nature of processes on a deeper level of organization. It is this level of organization we must keep in mind when speculating on "optimality," "capacity," "efficiency," "economy," "scope for activity," and, of course, "symmorphosis" in living systems.

## Acknowledgments

The original research referred to in this essay has been supported by the Fonds zur Förderung der wissenschaftlichen Forschung in Österreich, projects no. S-35, P-10113-BIO, P-10237-BIO. I thank my wife Joy for reading and improving the manuscript.

## Further reading

Brett, J. R. (1976) Scope for metabolism and growth of sockeye salmon, *Oncorhynchus nerka*, and some related energetics. *Journal of the Fisheries Research Board of Canada*, **33**, 307–13.

Elliott, J. M. (1976) The energetics of feeding, metabolism and growth of brown trout (*Salmo trutta* L.) in relation to body weight, water temperature and ration size. *Journal of Animal Ecology*, **45**, 923–48.

Fonds, M., Cronie, R., Vethaak, A. D. and Van der Puyl, P. (1992) Metabolism, food consumption and growth of plaice (*Pleuronectes platessa*) and flounder (*Platichthys flesus*) in relation to fish size and temperature. *Netherlands Journal of Sea Research*, **29**, 127–43.

Hammond, K. A., Konarzewski, M., Torres, R. M. and Diamond, J. M. (1995) Metabolic ceilings under a combination of peak energy demands. *Physiological Zoology*, **67**, 1479–506.

Hanel, R., Karjaleinen, J. and Wieser, W. (1996) Differentiation of swimming muscles, growth, and metabolic cost of growth, in larvae of whitefish (*Coregonus lavaretus*) at different temperatures. *Journal of Fish Biology*, **48**, 937–51.

Houde, E. D. (1989) Comparative growth, mortality, and energetics of marine fish larvae: temperature and implied latitudinal effects. *Fishery Bulletin of the United States*, **87**, 471–95.

Johnston, I. A. (1993) Temperature influences muscle differentiation and the relative timing of organogenesis in herring (*Clupea harengus*) larvae. *Marine Biology*, **116**, 363–79.

Randall, D. J. and Daxboeck, C. (1982) Cardiovascular changes in the rainbow trout (*Salmo gairdneri* Richardson) during exercise. *Canadian Journal of Zoology*, **60**, 1135–40.

Wieser, W. (1994) Cost of growth in cells and organisms: general rules and comparative aspects. *Biological Reviews*, **68**, 1–33.

Wieser, W. (1995) Energetics of fish larvae, the smallest vertebrates. *Acta Physiologica Scandinavica*, **154**, 279–90.

# 9

# Integrative systems for oxygen and fuel delivery

## 9.1 Overview

RICARDO MARTINEZ-RUIZ

This chapter deals with fuel utilization in mammals and optimality of design in the fuel supply system. The symmorphosis concept has been extensively studied by analyzing the oxygen flux across the respiratory system by Taylor, Weibel and their coworkers. From this monumental work it was established that the structures in the respiratory system were built to "meet but not exceed" the demands placed on them. The only exception was the lung, where an excess diffusing capacity relative to maximal oxygen consumption was found. The reason for this "uneconomical" design is not clear. Besides low malleability of the lung, the possibility of a built-in safety factor has been forwarded, with the argument that the lung is an interface organ with the environment.

However, complex systems tend to be multifunctional. It is clear that some steps of the respiratory cascade will be shared by substrate transport, for both oxygen and fuel will eventually be utilized in the mitochondria for energy production. Substrate pathways are more complex than that for oxygen: there is separation in time between uptake and utilization of substrates, hence the necessity of having carbohydrate and lipid stores. Substrate utilization was compared between "sedentary" goats and "athletic" dogs at different percentages of $\dot{V}_{O_2 max}$. It was possible to establish that even though the absolute levels of carbohydrates and lipid consumption were twice as high in the highly aerobic dogs compared to the goats, the relative percentage of the fuel composition was the same: lipid utilization reached its maximum between 40 and 60 percent $\dot{V}_{O_2 max}$, beyond which further substrate requirements were provided by carbohydrates. It was also shown that at the high levels of fuel consumption induced by running, both animals recruited the substrates mainly from intracellular stores; that is, muscles became practically closed systems during exercise. Dogs were also found to have twice as much in intra-muscular fuel stores, providing the muscle mitochondria with immediate energy supply for longer periods of time. This arrangement will circumvent the problem of substrate diffusion limitations at the sarcolemma.

Another way to evaluate the optimization of substrate utilization is to look at animals that have to endure prolonged fasts. Different strategies emerge whether the animal is lean or fat: during fasting, leaner animals tend to obtain a larger proportion of energy from proteins, which is often thought to reflect a poor survival potential. However, this is now viewed rather as a mechanism of optimizing fuel utilization in order to survive a prolonged fast. Its main goal is to prevent depletion of fuel stores before refeeding. To this end, fine tuning the right balance of fat and protein utilization throughout the period of fasting is critical.

In order to evaluate optimization of a system, it is useful to assess the output variable/input variable ratio. If the respiratory system of mammals was built "symmorphotically," you would expect the function/structure ratios to fall on the "line of symmorphosis"; that is, function is already maximized for the amount of structure available, and the more structure you have, the more function you will generate proportionally. As was pointed out before, the lung seems to maintain an excess diffusing capacity compared to $\dot{V}_{O_2\,max}$, except for the smallest mammals, where an extremely high specific $O_2$ consumption seems to push the system to the limit. However, this is difficult to prove because obtaining a blood sample in a 2 gram shrew for studying gas exchange will exsanguinate the animal. An alternative approach is to look at animals that have had strong selective pressures for optimizing their respiratory system: thoroughbred horses. It is found that the thoroughbred's respiratory system shows limitations for oxygen flux at $\dot{V}_{O_2\,max}$ at various simultaneous levels, so that the structures involved are being used at their maximum and there is no excess structure. In other words, the respiratory system seems to be optimized in this species in accordance with the symmorphosis concept. However, exploiting the system to the limit and thus foregoing any safety mechanisms has a high price: thoroughbred racehorses suffer from bleeding into their lungs as a result of the very high blood flows and blood pressures needed to achieve the high rates of oxygen transport – a good example of why safety factors are important.

## 9.2 Symmorphosis and the mammalian respiratory system: what is optimal design and does it exist?

### JAMES H. JONES

Optimality can be quantitatively defined as the maximization of output for a given input. Taylor and Weibel's hypothesized principle of symmorphosis states that an organism's systems should have no excess structure relative to their functional demands, a prediction of optimal design. If the mammalian respiratory system were designed according to symmorphosis, the ratios of functional to structural variables in the system should be constant and fall along a "line of symmorphosis" with a slope of unity when plotted against each other for different species (Figure 9.1). In contrast, if the system were not built according to symmorphosis, these ratios could fall virtually anywhere in the shaded region below the line of symmorphosis, indicating that some species have excess structure relative to the function achieved.

Taylor and Weibel compared the maximum $O_2$ transport function ($\dot{V}_{O_2 max}$) of the respiratory systems of different mammals with those systems' quantitative structure. They found that while some components (for example, mitochondrial volumes, $V_{mito}$, and muscle capillary lengths, $J_{cap}$, fell along a line of symmorphosis, the relationship between $\dot{V}_{O_2 max}$ and pulmonary diffusing capacity ($D_{L_{O_2}}$) clearly did not. When plotted on log-log coordinates, the relationship for $\dot{V}_{O_2 max}$ and $D_{L_{O_2}}$ was similar to the dashed line in Figure 9.1, suggesting that symmorphosis is not a general design principle for the respiratory systems of all mammals. However, this finding raises a fundamental question: is it possible for respiratory system design in *any* species of mammal to be optimized for $O_2$ transport?

The ideal animal in which to address this question would be that depicted by the solid circle in Figure 9.1 where the symmorphosis and $D_{L_{O_2}}$ lines converge. This smallest mammal, the 2 g Etruscan shrew (*Suncus etruscus*), is unfortunately too small for evaluation of physiolo-

241

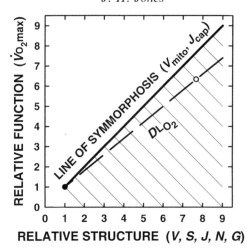

Figure 9.1. Possible relationships for various mammals between maximum respiratory function ($\dot{V}_{O_2\,max}$), in relative units, as a function of quantities of respiratory system structures ($V$, volumes; $S$, surface areas; $J$, lengths; $N$, numbers; $G$, conductances), in relative units. A constant ratio of function to structure among mammals would follow the line of symmorphosis, with a slope of unity, whereas points falling in the shaded region would have "excess" structure relative to function, and be suboptimal according to the principle of symmorphosis.

gical mechanisms of $O_2$ transfer. An alternative approach is to evaluate a species that shows extreme adaptation for respiratory function because of historically strong selective pressure. Thoroughbred racehorses are such animals, as horses have been selectively bred by humans for racing performance for several thousand years, and performance in flat races is highly correlated with peak aerobic power. The data in Figure 9.2 suggest that thoroughbreds may have reached a physiologic limitation to racing performance and a potentially optimized respiratory system. These graphs show the winning times over the past 70 to 150 years for the most elite British and American thoroughbred races. For all these races, the times appear to have reached asymptotic nadirs over the past two decades.

What determines if a respiratory system is optimized relative to its function and structure? The primary criterion is that each and every step in the $O_2$ transport system should contribute to the limitation to $O_2$ flux when the animal achieves $\dot{V}_{O_2\,max}$, as the structure at each step should be minimized relative to this maximum function ($O_2$ flux). If only a single step in the system were to show physiological evidence of limitation, this would suggest that excess structure existed in other steps rela-

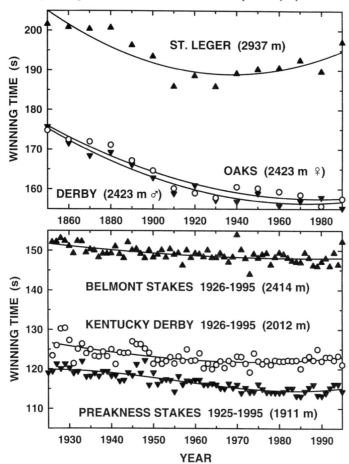

Figure 9.2.   Winning times for the most elite (Triple Crown) British (top) and American (bottom) thoroughbred horse flat races, with distances for each given in parentheses. For British races, data are 10-year averages back to 1850; for American races, annual times are shown back to 1925/26, the years in which the distances and handicaps were standardized.

tive to function at $\dot{V}_{O_2\,max}$. An optimized system need not have a high $O_2$ flux through it, merely a high flux relative to the quantity of structure that supports it.

   To determine if the thoroughbred respiratory system is optimized at $\dot{V}_{O_2\,max}$ I have conducted experiments during maximal treadmill exercise to determine the pattern of limitation to oxygen transport in their respiratory systems. The results of these experiments can be summarized in terms of the five structurally based components of the respiratory system.

## Mitochondrial oxidative capacity

Mitochondrial oxygen consumption at $\dot{V}_{O_2\,max}$ can be described as:

$$\dot{V}_{O_2\,max} = (\dot{V}_{O_2}/V_{mt}) \cdot V_{mt}$$

where $V_{mt}$ is the total volume of mitochondria in the exercising animal's skeletal muscle. This component of the respiratory system sets the demand for $O_2$ by reducing the partial pressure of $O_2$ ($P_{O_2}$) at the mitochondria, thereby causing $O_2$ to flow through the system. If this component of the respiratory system were limited at $\dot{V}_{O_2\,max}$, it would be impossible for other components of the system to reach their functional capacities and contribute to the systemic limitation, or for the system to be optimized.

When thoroughbred horses were run in hyperoxia (25 percent $O_2$) at sea level, their mass-specific $\dot{V}_{O_2\,max}$ ($\dot{V}_{O_2\,max}/M_b$, where $M_b$ is body mass) averaged 11 percent higher than when the same horses ran at sea level breathing air (21 percent $O_2$). This result unequivocally demonstrates that $\dot{V}_{O_2max}$ is limited by $O_2$ transport capacity, not mitochondrial oxidative capacity. A unique horse with a mitochondrial Complex I enzyme deficiency (a known mitochondrial oxidative limitation) had a $\dot{V}_{O_2\,max}/M_b$ that was only 20 percent of the normal horses, indicating an extreme mitochondrial limitation.

## Peripheral diffusion

Oxygen flux from muscle capillaries to mitochondria is described by the Bohr equation:

$$\dot{V}_{O_2\,max} = D_{T_{O_2}} \cdot \Delta P_{O_2}$$

where $D_{T_{O_2}}$ is the tissue diffusing capacity (conductance) and $\Delta P_{O_2}$ is the mean pressure difference from capillary to mitochondria that provides the potential for diffusion. If the structural components that determine $D_{T_{O_2}}$ were optimized, significant diffusion limitation should occur, the blood exiting the capillary should have a $P_{O_2}$ higher than that in the mitochondria (hypothesized to be near 0 torr), and the magnitude of the end-capillary (that is, venous) $P_{O_2}$ should be proportional to $\dot{V}_{O_2\,max}$. (Fick's law of diffusion requires that a greater $\Delta P_{O_2}$ would be required to generate a higher flux across the same $D_{T_{O_2}}$.)

Thoroughbreds experience a steady decline in mixed venous $P_{O_2}$ as they exercise at greater intensities from rest to $\dot{V}_{O_2\,max}$. However, when

$\dot{V}_{O_2 max}$ increased by 11 percent in hyperoxia, mixed venous $P_{O_2}$ increased by an average of 4 torr, consistent with the hypothesis that the animals experience peripheral tissue diffusion limitation at $\dot{V}_{O_2 max}$, and therefore must have a higher end-capillary $P_{O_2}$ in hyperoxia to achieve the higher $O_2$ flux that was observed. In contrast to these observations on normal horses, in the horse with a known mitochondrial oxidative limitation, mixed venous $P_{O_2}$ increased from rest to $\dot{V}_{O_2 max}$, indicating that its $O_2$ transport system was delivering $O_2$ to its mitochondria in excess of their capacity to use it.

### Circulatory convection

The structural components that describe convective $O_2$ flux, described by the Fick Principle, are as follows:

$$\dot{V}_{O_2 max} = \dot{Q}_T \cdot (Ca_{O_2} - C\bar{v}_{O_2})$$

where $\dot{Q}_T$ is cardiac output, and $Ca_{O_2}$ and $C\bar{v}_{O_2}$ are arterial and mixed venous $O_2$ concentrations, respectively. Circulatory function will be optimized when $\dot{Q}_T$, the product of heart rate and stroke volume (and the quotient of mean arterial pressure and total vascular resistance), is maximized relative to the size of the heart. Maximum heart rate is inversely related to body size, so increasing stroke volume, increasing driving pressure, and minimizing resistance will increase $\dot{Q}_T$. The arteriovenous $O_2$ concentration difference can be maximized if hemoglobin concentration is elevated until increased viscosity reduces $\dot{Q}_T$, and/or if $O_2$ extraction is maximized.

Trained thoroughbreds increase hematocrits from 45 percent at rest to 65 percent at $\dot{V}_{O_2 max}$, raising the $O_2$-carrying capacity of the blood. Very high $\dot{Q}_T / M_b$ is associated with mean systemic arterial pressures (240 versus 150 torr) and pulmonary arterial pressures ( > 100 versus 35 torr) at $\dot{V}_{O_2 max}$ that are much higher than in other mammals that have been studied. In conjunction with these unusually high vascular pressures, nearly all racehorses bleed from their lungs (exercise-induced pulmonary hemorrhage) when they run strenuously. The transmural pressure across the pulmonary capillaries exceeds their mechanical strength (stress failure), leading to structural failure and rupture. The cause of the high intravascular pressures at $\dot{V}_{O_2 max}$ does not appear to be high pulmonary vascular resistance, but rather, remarkably high (70 versus < 10 torr in most mammals) mean left atrial pressure, presumably required to rapidly fill the stiff left ventricle during diastole, forcing pressures up throughout the circulation.

## Pulmonary diffusion

Pulmonary gas exchange is similar to peripheral gas exchange in terms of the Bohr equation:

$$\dot{V}_{O_2 \max} = D_{L_{O_2}} \cdot \Delta P_{O_2}$$

where $D_{L_{O_2}}$ is pulmonary diffusing capacity (conductance) and $\Delta P_{O_2}$, the driving pressure for diffusion, is the mean difference between alveolar $P_{O_2}$ and capillary $P_{O_2}$. If maximum function were to be achieved for available structure, diffusion limitation should exist and there should be a difference between the alveolar and end-capillary $P_{O_2}$, reducing arterial $P_{O_2}$.

Thoroughbreds experience marked arterial hypoxemia during heavy exercise, from approximately 100 torr at rest to < 70 torr at $\dot{V}_{O_2 \max}$, with approximately 30 torr of diffusion limitation. This occurs despite the fact that Taylor and Weibel's initial studies showed that large mammals (for example, the horse) tend to have relatively more $D_{L_{O_2}}$ relative to their $\dot{V}_{O_2 \max}$ than do smaller mammals. When thoroughbred foals were raised in chronic cold and hypoxia at high altitude, they developed with 14 percent greater $D_{L_{O_2}}/M_b$ than sea-level controls, and had 15 percent greater $\dot{V}_{O_2 \max}/M_b$. Pulmonary diffusion limitation appears to contribute to limiting $O_2$ flux in exercising horses.

## Ventilatory convection

The ventilatory component of the system is described as follows:

$$\dot{V}_{O_2 \max} = \dot{V}_A \cdot (F_{I_{O_2}} - F_{\bar{E}_{O_2}})$$

where $\dot{V}_A$ is alveolar ventilation, and $F_{I_{O_2}}$ and $F_{\bar{E}_{O_2}}$ are inspired and mixed expired $O_2$ fractions, respectively. Ventilation becomes limited when it is unable to match the animal's rate of metabolic $CO_2$ production, resulting in increased alveolar (and arterial) $P_{CO_2}$, and a decrease in alveolar $P_{O_2}$; the animal hypoventilates.

Arterial $P_{CO_2}$ in thoroughbreds increases from 42 torr at rest to nearly 60 torr at $\dot{V}_{O_2 \max}$, contributing to the arterial hypoxemia previously described. Heavily exercising horses appear unable to increase $\dot{V}_A$, even when breathing gas containing 6 percent $CO_2$, the strongest ventilatory stimulant. It seems possible that thoroughbreds experience mechanical flow limitation in their airways as peak expiratory flows at $\dot{V}_{O_2\,max}$ exceed 100 liters s$^{-1}$.

The experiments described demonstrate that $\dot{V}_{O_2\,max}$ in thoroughbred horses is limited by all components of the $O_2$ transport system, not mitochondrial oxidative capacity:

(a) *peripheral diffusion*, as evidenced by the higher end-capillary $P_{O_2}$ required to increase $O_2$ diffusion into the muscle mitochondria in hyperoxia, consistent with peripheral diffusion limitation;
(b) *circulatory convection*, demonstrated by extremely high atrial and arterial pressures during exercise, and the catastrophic failure of pulmonary blood vessels that results;
(c) *pulmonary diffusion*, because of the large difference between alveolar and end-capillary $P_{O_2}$ due to diffusion limitation with resultant arterial hypoxemia, and the 15 percent increase in $\dot{V}_{O_2\,max}/M_b$ when $D_{L_{O_2}}/M_b$ was increased by 14 percent; and
(d) *ventilatory convection*, shown by increasing arterial and alveolar $P_{CO_2}$ during heavy exercise, and the apparent inability to increase alveolar ventilation to wash out $CO_2$, even when inspiring $CO_2$ while running. It is not currently known if this apparent ventilatory limitation is strictly mechanical.

Thoroughbred horses have been selectively bred for generations to enhance a phenotypic trait, running speed, that is associated with high aerobic capacity. Such highly directed selection appears to have optimized a complex integrated physiological system, the equine respiratory system. However, in considering this conclusion in conjunction with the allometric relationships that Taylor and Weibel found in testing their hypothesis of symmorphosis (Figure 9.1), a conundrum arises. The ratio of $\dot{V}_{O_2\,max}$ to $D_{L_{O_2}}$ in the horse is similar to that of the open circle shown on line A in Figure 9.1, well below the line of symmorphosis, suggesting it has "excess" $D_{L_{O_2}}$ relative to its $\dot{V}_{O_2\,max}$. The original premise on which Taylor and Weibel based their test of the hypothesis of symmorphosis – that structural variables should change in direct proportion to $\dot{V}_{O_2max}$ for symmorphosis to exist – is, therefore, called into question. Because the thoroughbred experiences severe pulmonary diffusion limitation despite having a greater $D_{L_{O_2}}$ relative to its $\dot{V}_{O_2\,max}$ than smaller mammals, we must conclude that structures in different mammals are not used identically when animals achieve their maximum functional outputs. Multiple transport steps in the system function in an integrated manner, so that changes in one transport step may affect adjacent steps. For instance, pressure heads for diffusion across a given structure (for example, the lung) are altered by changes in convection in adjacent transport steps. Therefore, optimization may occur, with each individual transport step contributing within the integrated system to a given animal's limitation, despite the fact that the ratio of function to structure for any given structure may be lower than in other species that are not optimized. Because the entire system produces the function, individual structural components can only be optimized in the context of the entire integrated system.

Ample evidence demonstrates that respiratory function in many species of mammals is not optimized, with some structures present in "excess" and contributing little to limiting the animal's maximum $O_2$ flux. However, as the racehorse demonstrates, given sufficient selective pressure, it is possible for a complex multicomponent integrated system, like the mammalian respiratory system, to become optimized for the function for which it has been selected.

## Further reading

Jones, J. H. and Lindstedt, S. L. (1993) Limits to maximal performance. *Annual Review of Physiology,* **55**, 547–69.

Jones, J. H. (1994) Circulatory function during exercise: integration of convection and diffusion. In *Comparative Vertebrate Exercise Physiology: Unifying Physiological Principles.* Ed. J. H. Jones. Academic Press, San Diego. *Advances in Veterinary Science and Comparative Medicine,* **38A**, 217–51.

Jones, J. H., Birks, E. K. and Pascoe, J. R. (1992) Factors limiting aerobic performance. In *The Vertebrate Oxygen Transport Cascade: From Atmosphere to Mitochondria,* pp. 169–78. Ed. J. E. P. W. Bicudo. CRC Press, Boca Raton.

Lindstedt, S. L. and Jones, J. H. (1987) Symmorphosis: the concept of optimal design. In *New Directions in Physiological Ecology*, pp. 289–309. Eds. M. Feder, A. F. Bennett, W. Burrgren and R. B. Huey. Cambridge University Press, New York.

Taylor, C. R., Karas, R. H., Weibel, E. R. and Hoppeler, H. (1987) Adaptive variation in the mammalian respiratory system in relation to energetic demand. *Respiration Physiology,* **69**, 1–127.

Weibel, E. R. and Taylor, C. R. (1981) Design of the mammalian respiratory system. *Respiration Physiology,* **44**, 1–164.

Weibel, E. R., Taylor, C. R. and Hoppeler, H. (1991) The concept of symmorphosis: a testable hypothesis of structure–function relationship. *Proceedings of the National Academy of Sciences USA,* **88**, 10357–61.

# 9.3  Adjusting maximal fuel delivery for differences in demand

## JEAN-MICHEL WEBER

Maximal energy consumption by muscle mitochondria varies considerably between animals, going from low rates in sedentary species such as goats and cows, to very high rates in the most athletic ones such as dogs and horses. Maximal supply rates of oxidative fuels and oxygen must consequently also vary to allow the maintenance of ATP homeostasis in working muscles of organisms with widely different exercise capacities. Endurance-adapted species which typically have a high aerobic capacity or $\dot{M}_{O_2\,max}$ have also adjusted the fuel supply pathways to their muscle mitochondria to allow very high rates of energy turnover during exercise (Weber 1988, 1992). This chapter discusses the metabolic changes necessary to achieve such high rates at the whole organism level. It focuses on the physiological measurements of oxidative substrate fluxes, in particular their partitioning into carbohydrate and lipid supply, and between vascular and intramuscular sources. The following chapter uses these data to correlate the functional parameters with the corresponding morphometric parameters of each fuel pathway in the same animals, using symmorphosis as a conceptual framework.

### Pathways for $O_2$ and fuel supply

Symmorphosis was originally proposed as a working hypothesis to investigate the quantitative structure–function relationships in functional systems made of a linear sequence of linked steps. Thus, the pathway for oxygen from atmospheric air to muscle mitochondria has been studied by comparing species with widely different aerobic capacities. This revealed that most of the steps of the oxygen transport pathway appear to be designed in proportion to $O_2$ demand in working muscles.

249

More recently, we have extended the investigation of mammalian locomotion energetics to include the supply of oxidative substrates, carbohydrates, and lipids, in addition to oxygen. The model describing these
connected pathways is now more complex as it combines serial and parallel steps (Figure 9.3). Accordingly, the analysis of structure–function
relations first requires the separate determination of maximal flux rates
from each oxidative substrate source available within the organism.
Oxidative fuel supply can be broken down into only four sources because
(1) two kinds of fuels are available (*lipids* and *carbohydrates*), and (2)
both can be stored either within the locomotory muscles themselves
(*intramuscular fuel stores*) or in other tissues from which they have to
be transported to muscles via the blood (*circulatory fuels*). Intramuscular
lipids are present as lipid droplets which are always found in close contact
with mitochondria (source 1; Figure 9.3) whereas intramuscular carbohydrates are deposited as glycogen granules (source 2); in contrast, circulatory fuels are provided to working muscles as plasma fatty acids
(source 3) or as plasma glucose (source 4). Circulatory fatty acids come
from the hydrolysis of triacylglycerol stores in adipose tissue or liver, and
circulatory glucose from gluconeogenesis and hepatic glycogen breakdown; during locomotion the direct supply from the gut is very small.
Therefore, under aerobic conditions, total muscle ATP turnover will
always be accounted for by the summed contributions from the oxidation
of these four distinct sources of fuel.

### Partitioning fuel supply

Do the relative contributions of these four sources to total metabolism
differ in animals with different aerobic capacity? The simplest scenario to
imagine is the parallel increase of maximal flux for each one of the four
fuels in exact proportion to changes in aerobic capacity. For example,
comparing a sedentary, low-aerobic goat ($\dot{M}_{O_2\,max} = 1\,ml\,O_2\,kg^{-1}\,s^{-1}$) to
a highly aerobic, endurance-adapted dog ($\dot{M}_{O_2\,max} = 2.2\,ml\,O_2\,kg^{-1}\,s^{-1}$)
could mean that maximal flux for each oxidative substrate is increased by
a factor of 2.2 in the dog. It would also mean that the *relative* contribution of each fuel source would be independent of $\dot{M}_{O_2\,max}$ when comparing these species at any given exercise intensity defined as percent
$\dot{M}_{O_2\,max}$. However, there are two main reasons to predict that a more
complex reorganization of fuel metabolism may occur in endurance-
adapted, high $\dot{M}_{O_2\,max}$ animals: (1) they may rely proportionately more
on lipids to take advantage of this very large energy source since well over

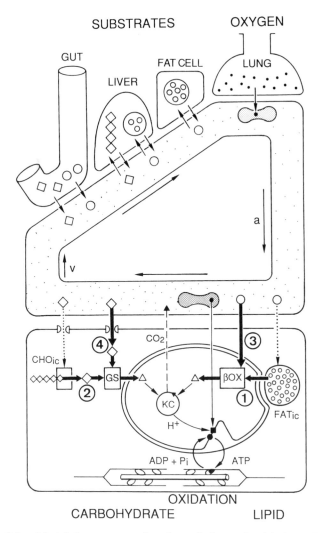

Figure 9.3. Model for structure–function relations of oxidative substrate and oxygen supply to locomotory muscle mitochondria from circulatory and intramuscular sources. Open circles represent fatty acids (bound to albumin in plasma and interstitial space and bound to fatty acid binding proteins in cytoplasm). Diamonds represent glucose, triangles represent acetyl CoA, and dots represent oxygen. The four pathways for oxidative fuel supporting ATP synthesis in muscle mitochondria are numbered: source 1 = intramuscular triacylglycerol (lipid droplets), source 2 = intramuscular glycogen granules, source 3 = circulatory fatty acids, and source 4 = circulatory glucose. GS = glycolysis, KC = Krebs cycle, and $\beta$OX = beta-oxidation. (Adapted from Weibel *et al.* 1996.)

90 percent of total fuel reserves are stored as fat in mammals; and (2) they may rely much more heavily on intramuscular fuel reserves than on circulatory substrates to avoid all the constraints associated with transport from storage tissues to muscle mitochondria via the circulation, such as the numerous transmembrane transport steps required for substrate supply, or the solubilization of lipids by specialized proteins necessary in the aqueous plasma and cytosol (Weber 1988).

We set out to measure maximal fluxes for the different fuels in two model species having widely different aerobic capacities: low-aerobic pigmy goats and endurance-adapted dogs (Roberts *et al.* 1996; Weber *et al.* 1996a,b). In these experiments, rates of total carbohydrate and total lipid oxidation were measured by indirect calorimetry, while continuous infusions of radiolabeled glucose and palmitate were used to quantify the contributions of circulating glucose and circulating fatty acids. Then, contributions from intramuscular glycogen and intramuscular triacylglycerol were simply calculated by subtraction. We found that maximal rates of lipid oxidation were reached at low exercise intensity ($\cong 40$ percent $\dot{M}_{O_2\,max}$) and that all the additional energy necessary to reach intensities above 40 percent $\dot{M}_{O_2\,max}$ was provided through the oxidation of carbohydrates, more specifically muscle glycogen. Contrary to expectation, total carbohydrate and total lipid oxidation were both increased in exact proportion to aerobic capacity when comparing dogs and goats at each exercise intensity. In other words, dogs running at the same percent $\dot{M}_{O_2\,max}$ as goats were oxidizing carbohydrates and lipids at 2.2 times the rate found for goats, and, therefore, the *relative* contributions of carbohydrate and lipid oxidation to total metabolism were identical for the two species. Examining these results in conjunction with data for other species from the literature revealed that these conclusions can be generalized to all mammals: highly aerobic animals do not favor the relative use of lipids, which was not what we had predicted (Roberts *et al.* 1996).

Figure 9.4 shows the partitioning of fuel supply between vascular sources, and intramuscular stores at different exercise intensities. For both fuels, vascular supply reaches its maximum at low exercise intensities, and all the additional energy required at higher intensities is drawn from intramuscular stores. Maximal rates of circulatory glucose oxidation (Weber *et al.* 1996a) and circulatory fatty acid oxidation (Weber *et al.* 1996b) were somewhat higher in endurance-adapted dogs than in their sedentary counterparts, but not in proportion to their respective aerobic capacities. Therefore, the relative importance of circulatory fuel oxida-

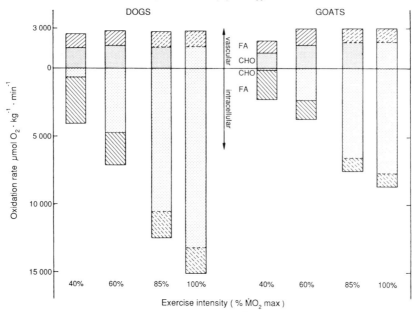

Figure 9.4. Rates of oxidation of glucose (CHO) and fatty acids (FA) at different exercise intensities partitioned with respect to the supply pathways from vascular and intramuscular substrate pools. Oxidation rate of vascular substrates is plotted above the baseline, that of intramuscular substrates below the baseline. Calculated after data from Roberts *et al.* (1996) and Weber *et al.* (1996a,b). The partitioning of fatty acid supply at 85 percent and 100 percent $M_{O_2\,max}$ is extrapolated and shown with broken line and different shading. (From Weibel *et al.* 1996.)

tion to total metabolism was much smaller in the athletic dogs than in the low-aerobic goats. In the dogs, this relative shortfall in energy provision from the circulation was compensated for by a much higher reliance on muscle glycogen and muscle lipid reserves.

## Conclusions

We found that highly aerobic mammals have adjusted their oxidative fuel supply system for exceptionally high energetic demands by relying proportionately less on circulatory substrates and relatively more on intramuscular fuels. Extreme adaptation for endurance means that locomotory muscles become practically closed systems for oxidative fuels during exercise, and use periods of rest to replenish their internal fuel reserves via low circulatory substrate fluxes. The direct proximity of intramuscular lipid droplets to muscle mitochondria emphasizes the notion that this strategy is necessary to circumvent the limitations associated with fuel transport across membranes and with solubilization of lipids in aqueous biological fluids. Using symmorphosis to analyze the structural design of muscle, the following chapter examines the key structural elements involved in fuel transport and correlates them with the maximal fluxes from intramuscular and circulatory substrate sources discussed here.

## Further reading

Roberts, T. J., Weber, J.-M., Hoppeler, H., Weibel, E. R. and Taylor, C. R. (1996) Design of the oxygen and substrate pathways: II. Defining the upper limits of carbohydrate and fat oxidation. *J. Exp. Biol.*, **199**, 1650–8.

Weber, J.-M. (1988) Design of exogenous fuel supply systems: adaptive strategies for endurance locomotion. *Can. J. Zool.*, **66**, 1116–21.

Weber, J.-M. (1992) Pathways for oxidative fuel provision to working muscles: Ecological consequences of maximal supply limitations. *Experientia*, **48**, 557–64.

Weber, J.-M., Roberts, T. J., Vock, R., Weibel, E. R. and Taylor, C. R. (1996a) Design of the oxygen and substrate pathways: III. Partitioning energy provision from carbohydrates. *J. Exp. Biol.*, **199**, 1659–66.

Weber, J.-M., Brichon, G., Zwingelstein, G., McClelland, G., Saucedo, C., Weibel, E. R. and Taylor, C. R. (1996b) Design of the oxygen and substrate pathways: IV. Partitioning energy provision from fatty acids. *J. Exp. Biol.*, **199**, 1667–74.

Weibel, E. R., Taylor, C. R., Weber, J.-M., Vock, R., Roberts, T. J. and Hoppeler, H. (1996) Design of the oxygen and substrate pathways: VII. Different structural limits for $O_2$ and substrate supply to muscle mitochondria. *J. Exp. Biol.*, **199**, 1699–1709.

## 9.4 The converging pathways for oxygen and substrates in muscle mitochondria

### HANS HOPPELER

The principle of symmorphosis was set up to complement the principle of homeostasis; the latter acting on functional regulation, the former acting on morphogenesis (Taylor and Weibel 1981). Homeostasis can be studied at work; symmorphosis cannot be observed operating in real time. The process of adjusting structural capacity to functional demand is the result of natural selection (for example, Pronghorn antelopes) or of selective breeding (for example, thoroughbred racehorses).

#### The respiratory cascade

To test the concept of symmorphosis for the respiratory system we first analyzed the pathway for oxygen from the lungs to skeletal muscle mitochondria using allometric and adaptive variation in $\dot{V}_{O_2 max}$ as an experimental tool. Symmorphosis was found to prevail at all levels of the respiratory system except for the lungs. Because of the limited malleability of the lungs it was hypothesized that this might put small and active animals at a disadvantage with regard to adaptability of "downstream" elements of the respiratory cascade, as well as with regard to aerobic performance in a hypoxic environment. Alternatively, it was proposed to consider the variable overdesign of the lungs as a "safety factor."

The analysis of the respiratory system was carried out under the assumptions (a) that oxygen transfer served a vital and dominant function; (b) that this function reached a measurable and well-defined limit ($\dot{V}_{O_2 max}$); (c) that this limit varied widely among species; and (d) that the subservience of structural entities defining functional capacities was adequately established (Weibel, Taylor and Hoppeler 1992).

A major critique of the analysis of the respiratory cascade as a test case for symmorphosis entailed the arbitrary selection of a single function,

255

*oxygen flux rate*, as the sole variable for which structural capacities should be matched on all levels of the system, for it is quite clear that cellular oxidation depends as much on substrate as on oxygen availability. Common transfer steps could thus be designed to serve either oxygen *or* substrate transfer functions. We have therefore extended our earlier studies to include the pathways supplying substrates to mitochondria. The oxygen and substrate pathways converge in the mitochondria and there are several steps where the same structures transport both. We have previously concluded that these structures are quantitatively matched to the maximal rate of oxygen utilization. We now take into consideration that the same structures also have to ensure delivery of substrates at matching rates. In an extended model of oxygen *and* substrate transport to muscle mitochondria, we define the relevant structural and functional variables and predict that the structural components will be matched to the functional capacities.

### Establishing the pathways for oxygen and substrate supply to mitochondria

Oxygen follows a single pathway from lungs to skeletal muscle mitochondria and steady-state flux rates are established after a few minutes. Oxygen stores in myoglobin are of little importance in terrestrial mammals, and the highest rate of mitochondrial oxygen consumption coincides with the highest rate of oxygen uptake and transport by lung and circulation. The substrate pathways are more complex. There are organismic substrate stores in the liver (lipids and carbohydrates) and in adipose tissue (lipids only). Additionally, the muscle cell itself stores varying amounts of substrates, carbohydrates in the form of glycogen granules, and lipids in the form of small droplets of triacylglycerol (Figure 9.5). Moreover, maximal rates of substrate uptake by the gut and substrate utilization in muscle are separated in time and regulated independently.

Maximum substrate oxidation occurs during exercise when substrate uptake is minimal. When looking at substrate supply to the mitochondria there are four pathways by which substrates can be delivered during exercise (see Weber, Chapter 9.3). Glucose and lipids mobilized from body stores are transported in the circulating plasma and reach mitochondria after traversing several barriers: capillary endothelium, interstitial space, sarcolemma, and cytosol. Some of these transport steps may involve specific transporters (for example, the membrane-bound glucose

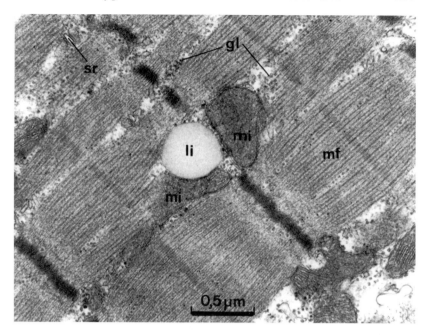

Figure 9.5. Electron micrograph of dog triceps muscle showing the intracellular substrate stores glycogen (gl) and lipid droplets (li). Lipid droplets are always found tightly associated with mitochondria (mi) whereas glycogen is also found interspersed between myofilaments (mf); sr denotes sarcoplasmic reticulum. Scale marker is 0.5 μm.

transporters GLUT-1 and GLUT-4). The pathways from intracellular stores for carbohydrates and lipids are short, providing mitochondria with almost immediate access to these substrates. The flux rates from intravascular pools may be limited by structural properties of the intervening barriers and spaces while both the location and extent of the intracellular stores are important design parameters possibly affecting substrate availability to mitochondria (Vock *et al.* 1996a,b).

### Experimental design

To vary the capacity for aerobic metabolism we compared "athletic" dogs to "sedentary" goats of the same body mass whereby the $\dot{V}_{O_2\,max}/M_b$ of dogs was more than two-fold that of goats. In a series of physiological experiments (see Weber, Chapter 9.3) the relative rates of carbohydrate and lipid oxidation for different work intensities was established by indirect calorimetry, and the rates of oxidation of circulatory glucose and fatty acids were determined. Subtracting these rates from the total oxidation rates of the respective substrates yielded the rate of oxidation of intracellular stores of glucose and lipids. After completion of the physiological studies the animals were sacrificed and representative samples of the entire musculature were collected for morphometric and biochemical analysis.

### Physiological results summarized

The physiological results discussed in the companion paper (see Weber, Chapter 9.3) can be summarized as follows. As exercise intensity goes up from rest, the rate of lipid oxidation increases to reach a maximum at about 60 percent of $\dot{V}_{O_2 max}$ in both species when two-thirds of the energy is supplied by fat oxidation. Further increases in exercise intensity are fuelled entirely by carbohydrates which provide 80 percent of the energy at 85 percent of $\dot{V}_{O_2 max}$. The rates at which glucose and free fatty acids are supplied from the circulation increase with exercise intensity and reach a maximum between 40 and 60 percent of $\dot{V}_{O_2 max}$. Highly aerobic dogs can use about the same amount of circulatory glucose and fatty acids as goats; as a consequence of their higher $\dot{V}_{O_2 max}$ the aerobic species must rely more heavily on intracellular fuel reserves, both of carbohydrates and lipids.

### Glucose pathways

#### *Glucose transfer from capillaries*

To arrive in muscle cells from plasma, glucose molecules must pass the endothelial barrier by restricted diffusion through the small pore system. Here the relevant structural parameter is the overall size and number of the pores; the relevant functional parameter is the concentration difference between plasma and interstitial space. The sarcolemma is reached by free diffusion, then glucose is transferred into the interior of the cell by facilitative glucose transporters. Calculations taking into account the endothelial junction length, the permeability characteristics of the pore system, as well as the transendothelial concentration gradient, show that the capillary endothelium offers but a weak resistance to the diffusion of glucose from the plasma to the interstitial space. Likewise, the interstitial space (partly because it is so narrow) does not seem to be a site of major resistance to glucose flux to the sarcolemma. The morphometric analysis shows the surface area of sarcolemma to be invariant between dogs and goats (Figure 9.6a,b). The facilitative transporter GLUT-4 is known to be recruited to the sarcolemma at moderate exercise intensities and we measured the same maximal glucose flux density of about $2.5 \, \mu\text{mol} \, \text{min}^{-1} \, \text{m}^{-2}$ between 40 and 60 percent of $\dot{V}_{\text{O}_2 \, \text{max}}$ in dogs and goats. It is therefore likely that the sarcolemmal transporters could become saturated at these exercise intensities and that, as a result, the sarcolemma limits glucose uptake by the muscle cell.

### Intracellular carbohydrate

Glycogen occurs in uniform granules about 10 nm in diameter, some of them dispersed throughout the muscle cell, some packed around mitochondria (Figure 9.5). The essential finding is that dogs store as much as four times more glycogen in their muscles than goats (Figure 9.6a). As the rate at which dogs, use glycogen at 85 percent of $\dot{V}_{\text{O}_2 \, \text{max}}$ is only 1.6-fold that of goats, it follows that the dogs' intracellular carbohydrate reserves last more than twice as long as those of goats. It has to be considered further that glycogen may also be used at very high rates when glycolysis provides ATP during high-intensity anaerobic work.

We note that only the recruitment of glucose from intracellular glycogen stores can be increased to cover the higher fuel needs during high-intensity work. Dogs store more glycogen in their muscles than goats, and this allows them both to run faster and for longer time periods. Alternatively, they can use their glycogen reserves during protracted sprints.

### Lipid pathways

#### Lipid transfer from capillaries

In the plasma and in the interstitial space the hydrophobic fatty acids are solubilized by binding to albumin. Due to a higher fatty acid binding capacity of their albumin, dogs have twice the free fatty acid concentration in their plasma than that for goats. The mechanism of transfer of fatty acids across the two barriers, endothelium and sarcolemma, is disputed; it is unclear whether a specific fatty acid transporter exists in the sarcolemma. In the interior of the muscle cell, fatty acids are transported bound to fatty acid binding proteins. As for carbohydrates, we found

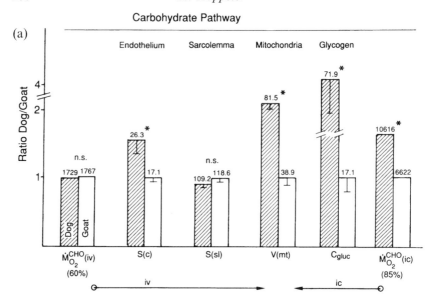

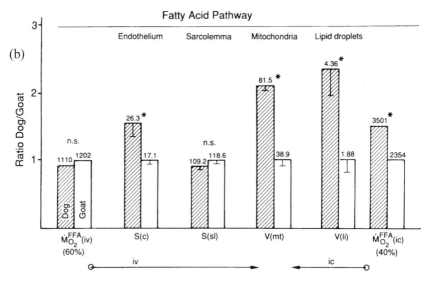

Figure 9.6. Comparison of physiological and morphometric data of (a) the carbohydrate and (b) the fatty acid pathways; $\dot{M}_{O_2}$ = maximal rate of oxidation, iv = supply from intravascular; ic = supply from intracellular; $S(c)$ = surface area of capillaries; $S(sl)$ = surface of sarcolemma; $V(mt)$ = volume of mitochondria; $C_{gluc}$ = concentration of glycogen; $V(li)$ = volume of lipid droplets.

dogs to extract similar quantities of fatty acids from the circulation as goats, and that this transport was not upregulated at higher work intensities. The net result of our calculations shows that both species extract only some 3.5 percent of their plasma fatty acid pool during each transit time. This extraction fraction is similar to that for glucose. The small quantity of substrate uptake from the circulation argues for a design limitation of substrate transfer from the circulation, possibly at the level of the sarcolemma.

### *Intracellular lipid*

Triacylglycerol droplets tightly associated with mitochondria represent the intracellular lipid stores of muscle cells (Figure 9.5). We found that when fat oxidation is maximal (at 40 percent of $\dot{V}_{O_2\,max}$) two-thirds of the fatty acids consumed are drawn from intracellular deposits with maximal rates being 50 percent higher in dogs than in goats. Similar to glycogen, we found intracellular lipid stores in dogs to be more than twice as big as those in goats (Figure 9.6b). Lipid stores therefore last 50 percent longer in dogs than in goats; they also last twice as long as the carbohydrate stores. Interestingly the lipid–mitochondria contact surface is four times larger in dogs than in goats, affording the athletic species with considerable structural redundancy for the transfer of fatty acids to the mitochondria.

We note that the supply of fatty acids from intravascular sources as well as from intracellular stores is limited. Possible sites of limitation are either the transfer steps of fatty acids to mitochondria (including the carnitine shuttle) or their capacity for fatty acid oxidation.

### Conclusions

The present study confirms and extends previous conclusions on the design of the pathway for oxygen: the volume of mitochondria, the size of the microvasculature and the erythrocyte content are coadjusted to support maximal oxygen flux. The regulation of oxygen supply is achieved by upregulating vascular perfusion. In contrast, upregulation of substrate supply depends completely on an increased utilization of intracellular substrate deposits. The structures of the supply pathway from blood to mitochondria are inadequate to supply glucose and fatty acids at the rates required during exercise intensities above 40 percent of $\dot{V}_{O_2\,max}$. They are designed primarily for the steady supply of $O_2$ at maximal rates of oxidation. Circulatory substrate supply occurs at low rates during periods of rest or low activity to build up intracellular substrates stores that can be exploited during exercise. The principle of symmorphosis was useful in structuring the analysis of the complex system of cellular aerobic metabolism, allowing us to pinpoint the sarcolemma as a likely candidate for limiting substrate flux into working muscle cells.

## Acknowledgments

This work was supported by grants from the Swiss National Science Foundation (31-30946.91), the Maurice E. Müller Foundation, Berne, the U.S. National Science Foundation (IBN 89-18371), the U.S. National Institutes of Health (AR 18140), and the Natural Sciences and Engineering Research Council of Canada.

## Further reading

Taylor, C. R. and Weibel, E. R. (1981) Design of the mammalian respiratory system. I: Problem and strategy. *Respir. Physiol.*, **44**, 1–10.
Vock, R., Weibel, E. R., Hoppeler, H., Ordway, G., Weber, J.-M. and Taylor, C. R. (1996a) Design of the oxygen and substrate pathways. V. Structural basis of vascular substrate supply to muscle cells. *J. Exp. Biol.*, **199**(8), 1675–88.
Vock, R., Hoppeler, H., Claassen, H., Wu, D. X. Y., Billeter, R., Weber, J.-M., Taylor, C. R. and Weibel, E. R. (1996b) Design of the oxygen and substrate pathways. VI. Structural basis of intracellular substrate supply to mitochondria in muscle cells. *J. Exp. Biol.*, **199**(8), 1689–97.
Weibel E. R., Taylor, C. R. and Hoppeler, H. (1992) Variations in function and design: Testing symmorphosis in the respiratory system. *Respir. Physiol.*, **87**, 325–48.
Weibel, E. R., Taylor, C. R., Weber, J.-M., Vock, R., Roberts, T. J. and Hoppeler, H. (1996) Design of the oxygen and substrate pathways: VII. Different structural limits for $O_2$ and substrate supply to muscle mitochondria. *J. Exp. Biol.*, **199**, 1699–709.

# 9.5  Fuel specialists for endurance

YVON LE MAHO and MOHAMED BNOUHAM

During periods of unfavorable weather, wild animals may face severe restriction in food resources while their energy requirements are higher due to cold exposure. In addition, many wild animals spontaneously undergo long-term fasts in relation to periods with a predictable drop in food resources or in relation to specific activities such as migration, moult, or breeding (Mrosovsky and Sherry 1980). Accordingly, animal survival and/or success in specific activities partly depend on the optimization of body fuel storage; that is, body fuel available in relation to predictable duration of food restriction. Survival and/or success in activity also depend on the optimization of the utilization of body fuel reserves. To endure a long fast, minimizing fat mobilization is not the only goal. According to Cahill (1976), a key factor for survival to a long fast is considered to be the ability to minimize protein loss and to prolong this so-called "protein sparing." Proteins are not stored raw materials. They are made to serve specific functions, particularly structural, muscular, and enzymatic functions. Burning proteins obviously reduce the capacity of these functions. To our knowledge, the torpid bear is the only animal that is able to save body protein totally without any food (Nelson 1973). For all the other animals, as well as humans, a decrease in body protein utilization is a major factor in withstanding a long fast.

Accordingly, the further rise in protein utilization that occurs during a prolonged fast has for many years been considered as an irreversible and pathologic process, the so-called "premortal rise in nitrogen excretion" (Grande 1964). Similarly, the higher protein loss of a lean animal during fasting compared to an obese animal appeared to be the result of a poorer ability to reach an effective protein sparing (Goodman et al. 1980).

As we will see in this part of the chapter, based on the study of wild animals that are professional fasters, the further rise in protein utilization

263

and the poorer ability of a lean animal to spare body protein instead appear as an optimization of body fuel utilization in order to survive a long fast.

## Optimization of body fuel storage

Optimization of body fuel storage is demonstrated by the fact that fat storage is roughly correlated with the duration of impending starvation and activity. Fat storage varies from 5–6 percent of body mass in birds before nocturnal food deprivation to 40 percent for over-sea migrant species (Cherel, Robin and Le Maho 1988). That no more fat than requested is stored may be explained by several limiting factors. To insure locomotion at maximum efficiency, more fat means more power and therefore more muscle. However, the increase in body mass that results from more fat and muscle means a higher cost of locomotion and possibly a more reduced ability of movements to escape from predation; on the other hand, the amount of fat that is stored is important in itself as a fuel but also because, as indicated below, it enables better protein utilization during a long-term fast (Le Maho et al. 1988).

Not only the amount of fat that is stored may be optimized before a period of predictable food shortage. Fatty acid composition may also be optimized. This has been illustrated for Yellow pine chipmunks, *Eutamias amoenus*, prior to hibernation. Their ability to lower body temperature, which is essential in order to endure a long-term reduction in food intake, depends on the selective intake of polyunsaturated fatty acids (Geiser and Kenagy 1987).

In addition, although it is commonly assumed that fuel storage essentially refers to fat storage (that is, changes in body protein are often neglected), it now appears that at least a part of body protein may be stored as a fuel. This is illustrated by the king penguin: due to the higher protein requirements during the molt (protein is required in the production of the new feathers), much more body protein is accumulated before the fast that is associated with molt, than before the fast that coincides with breeding (Cherel et al. 1994).

## Optimization of body fuel utilization during fasting

To determine how the utilization of body fuel may enable the endurance of a long fast, it is obviously essential to take into account the physiological limits for the depletion of both lipids stores and body protein.

Optimization of body fuel utilization therefore implies that the respective breakdown of fat stores and body proteins are adjusted so that a critical depletion in both body fuels would be reached at about the same time. Moreover, for an animal that still has the possibility to search for food, (for example, by migration further south or interruption of the spontaneous fast that is associated with breeding), optimization of body fuel utilization also implies that refeeding is triggered before a critical depletion in body fuels; that is, while body fuel reserves are still available to provide the energy requirements to initiate foraging.

### Physiological limits

A key point is that the physiological limits for the depletion of body fat stores differ from those for body protein. Although life is not possible without fat, very little fat can be afforded; for example, many tropical animals live with only about 5 percent of their body as fat. In contrast to how easily fat can be stored or depleted, death usually occurs when about 50–60 percent of body protein has disappeared. This explains why protein sparing is a major adaptation to enable endurance of a long fast.

Moreover, based on data from the literature, those limiting fuels for survival during a long-term fast differ between lean and obese individuals. Sudden deaths in very obese humans undergoing long-term fasting as a treatment of obesity may be attributed to the critical loss of about half of body protein while fat is still far in excess (Le Maho *et al.* 1988). In contrast, survival appears to be limited by the availability of fat in the lean individual (Cherel *et al.* 1992).

### Interaction between fat stores and body protein breakdown

#### Role of initial fatness

In animals and humans, about 20 percent of the energy required is derived from body protein in the lean individual, as compared to 5–10 percent in the fat one (Le Maho *et al.* 1988). This proportion is maintained at a steady value until a further rise in nitrogen utilization. In an emperor penguin this occurs after 3–4 months (Robin *et al.* 1988), although 80 percent of fat reserves disappear during that time (Figure 9.7). Thus, protein loss does not merely change in relation to fat availability.

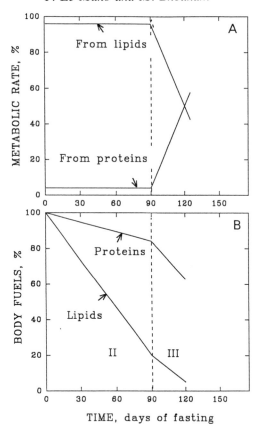

Figure 9.7. (A) Schematic change in the relative contribution of lipid reserves versus body proteins to energy expenditure during the course of long-term fast in an emperor penguin. (B) Change in lipid reserves versus body protein in relation to their initial level.

Considering protein sparing as a key factor in enduring a long fast, the larger proportion of the energy that is derived from body protein in a lean animal, could *a priori* be considered to reflect a poorer ability for survival. However, if a lean animal were to mobilize its fat stores as much as a fat animal, it would reach a critical level in fat depletion more rapidly while body protein is still readily available. Thus, that effectiveness in body protein sparing depends on the initial amount of body fat corresponds to an optimization of the relative contribution of lipid and protein to energy expenditure. Accordingly, a lethal level in lipid loss will not be reached in an initially lean animal while body protein is still available.

## The further rise in body protein breakdown

Still in relation to protein sparing as a major process to endure a long fast, the further rise in protein breakdown could be considered as a degenerative process. However, wild animals that are professional fasters, like penguins, may spontaneously enter into this metabolic stage, which is still reversible, as was also found for laboratory animals. Moreover, the further rise in protein breakdown is associated in wild animals like penguins with some "signal" which triggers refeeding (Le Maho *et al.* 1988), while fat reserves are still sufficient to cover the energy requirements to search for food. Since the part of basal metabolic rate that is derived from fat stores is decreasing at that time, the further rise in body protein therefore appears as an optimization of body fuel utilization, a critical depletion in fat stores being anticipated.

## Conclusion

In conclusion, therefore, there is some regulatory process at the onset of a fast which induces an adjustment of the relative contribution of fat stores and body protein to energy expenditure in relation to their respective level. However, there are some limits in this adjustment of body fuel utilization, when the animal is either very lean or very fat. There is no way during a prolonged fast that an extremely lean animal can avoid reaching a lethal level in lipid reserves while body protein is still readily available.

Similarly, in an extremely obese individual, initial adiposity is such that a lethal level in body protein will be reached during fasting even though huge fat stores are still available. Optimization for endurance of a long fast is therefore primarily based on the right balance and amounts in the respective accumulation of body protein and fat reserves before a predictable enforced fast or a spontaneous fast.

## Further reading

Cahill, G. F., Jr. (1976) Starvation in man. *Clin. Endocrinol. Metab.*, **5**, 397-415.

Cherel, Y., Robin, J.-P. and Le Maho, Y. (1988) Physiology and biochemistry of long-term fasting in birds. *Can. J. Zool.*, **66**, 159-66.

Cherel, Y., Robin, J.-P., Heitz, A., Calgari, C. and Le Maho, Y. (1992) Relationships between lipid availability and protein utilization during prolonged fasting. *J. Comp. Physiol. B*, **162**, 305-13.

Cherel, Y., Gilles, J., Handrich, Y. and Le Maho, Y. (1994) Nutrient reserve dynamics and energetics during long-term fasting in the King Penguin (*Aptenodytes patagonicus*). *J. Zool. Lond.*, **234**, 1–12.

Geiser, F. and Kenagy, G. J. (1987) Polyunsaturated lipid diet lengthens torpor and reduces body temperature in a hibernator. *Am. J. Physiol.*, **252** (*Regulatory Integrative Comp. Physiol.*, **21**), R897–R901.

Goodman, M. N., Larsen, P. R., Kaplan, M. M., Aoki, T. T., Young, V. R. and Ruderman, N. B. (1980) Starvation in the rat. II. Effect of age and obesity on protein sparing and fuel metabolism. *Am. J. Physiol.*, **239** (*Endocrinol. Metab.*, **2**), E277–86.

Grande, F. (1964) Man under caloric deficiency. In *Handbook of Physiology, Section 4: Adaptation to the Environment*, pp. 911–37. American Physiological Society, Washington, DC.

Le Maho, Y., Robin, J.-P. and Cherel, Y. (1988) Starvation as a treatment for obesity: the need to conserve body protein. *News Physiol. Sci.*, **3**, 21–4.

Mrosovsky, N. and Sherry, D. F. (1980) Animal anorexias. *Science* (Washington, DC), **207**, 837–42.

Nelson, R. A. (1973) Winter sleep in the black bear. A physiologic and metabolic marvel. *Mayo Clin. Proc.*, **48**, 733–37.

Robin, J. P., Frain, M., Sardet, C., Groscolas, R. and Le Maho, Y. (1988) Protein and lipid utilization during long-term fasting in emperor penguins. *Am. J. Physiol.*, **254**, R61–R68.

# 10

# Design of nervous systems

## 10.1  Overview

### RICHARD D. KEYNES

Nervous systems are designed to enable organisms to react appropriately and rapidly to changes in their external environment, and to regulate their internal environment closely. This requires the provision of sense organs specialized for vision and the detection of radiation in the optical spectrum, hearing and the detection of sound both in air and in water, the detection of gravitational pull and the forces that act when direction of movement is changed, the detection of electrical currents in water, mechanoreceptors and proprioreceptors, temperature receptors, and sensitive chemoreceptors. The information derived from these various sensors has to be analyzed and coded, and transmitted reliably to other parts of the nervous system to elicit a suitable mechanical response, the control of which may involve complicated feedback loops with receptors.

One feature of all nervous systems throughout the Animal Kingdom is that although the operation of sense organs must necessarily involve a graded response at the detector level, the transmission of information by peripheral nerve fibers is invariably digitalized. It may therefore be argued that the two principal operative factors that governed the original evolution of nerve fibers was their reliability and their conduction velocity. When vertebrates appeared on the stage, the introduction of myelinization that speeded up conduction and greatly reduced the bulk of peripheral nerve trunks played an important part in facilitating the development of highly centralized and well-coordinated nervous systems that invertebrates had lacked, and at the same time enabled the size of individual animals to increase. As far as symmorphosis is concerned, myelinization also carries benefits in body weight and consumption of energy, but except in some special cases it would seem likely that an improvement in operating efficiency that is not always readily quantifiable is the main advantage that it bestows.

There is not space in this volume to examine the optimization of design in nervous systems in general, but research that has been done on the visual systems of two very different animals provides admirable examples of the manner in which symmorphosis may operate. Dr Laughlin's elegant analysis of the performance of the compound eye of the blowfly demonstrates how remarkably well it is designed within the basic constraints imposed by its phylogeny, but he concludes that the value of good vision must ultimately be related to fitness, and that the superior optical design of the "simple" eye of vertebrates makes it ultimately the winner. Is it an accident that the largest and most highly developed invertebrates of the deep oceans, the cephalopods, have simple rather than compound eyes?

Professor Nevo gives an interestingly contrasted account of the evolution of the eye of the blind mole rat *Spalax*, in which a thorough adaptation to life underground has led to a complete loss of visual function but retention of a retinal capacity for photoperiodic perception. At the same time, a variety of nonvisual sensory systems for communication and orientation have been highly developed, and an extension of the somatosensory cortex of the brain has replaced the occipital visual cortex. Again it would seem that fitness is an important constraint to be taken into account in considering development along such a different pathway.

### Further reading

Abbott, N. J., Williamson, R. and Maddock, L. (1995) *Cephalopod Neurobiology. Neuroscience Studies in Squid, Octopus, and Cuttlefish.* Oxford University Press, Oxford, UK.

Aidley, D. J. and Stanfield, P. R. (1996) *Ion Channels, Molecules in Action.* Cambridge University Press, Cambridge, UK.

Bolis L., Keynes R. D. and Maddrell, S. H. P. (1984) *Comparative Physiology of Sensory Systems.* Cambridge University Press, Cambridge, UK.

Shepherd, G. M. (1988) *Neurobiology*, 2nd edn. Oxford University Press, Oxford, UK.

# 10.2   The design of peripheral nerve fibers

## RICHARD D. KEYNES

One of the essentials for an efficient nervous system was established early
on by the evolution of voltage-gated ion channels selective for sodium
and potassium which, as was first shown by Hodgkin and Huxley (1952)
in a classic paper, are responsible for the mechanism of conduction of
impulses in peripheral nerves throughout the Animal Kingdom. The
sodium channels operate in an all-or-none fashion, and thus permit the
transmission of information in the essentially digital form of discrete
impulses, which – as we are reminded every day – is the most efficient
way of handling data.

### The sodium channel

As shown in Figure 10.1, the sodium channel is a large membrane protein
comprising a chain of some 1820 amino acids arranged with four internally homologous domains, each built up of six membrane-spanning $\alpha$-
helices. The most closely conserved parts of the structure are the S4 $\alpha$-
helices that act as voltage-sensors, and respond to a depolarization of the
membrane by briefly opening the channel. Their screw-helical structure,
embodying a sequence of positively charged arginine or lysine residues in
every third position, separated by hydrophobic residues, enables them to
move in a series of discrete steps, each involving the effective transfer of
one elementary charge across the membrane. In domains I, II, and III the
S4 units all carry five positive charges, while in domain IV there are
throughout the Animal Kingdom, all the way from the fly *Drosophila*
to the SkM1 gene of human muscle fibers, eight positively charged residues spaced identically. So it would seem that an optimum design for the
part of the building block responsible for the activation of voltage-gated
ion channels was evolved very soon, and has not been improved upon for

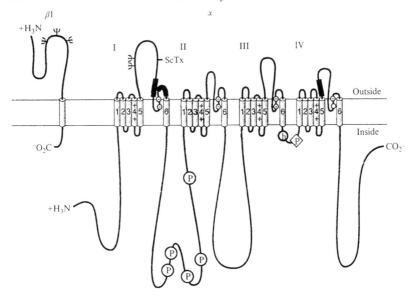

Figure 10.1.   Primary structures of the $\alpha$- and $\beta_1$-subunits of a rat brain sodium channel, shown as transmembrane folding diagrams. Bold lines are polypeptide chains. Cylinders represent $\alpha$-helices S1 to S6 traversing the membrane. For an explanation of other details, see Figure 1 of Keynes (1994).

hundreds of millions of years. At the same time, there are isoforms affecting the structure of other parts of the channel, and that regulate the rate and completeness of inactivation, which may come into play at different stages in development.

## Conduction velocity and axon size

The second most important characteristic of an efficient peripheral nerve fiber is that both the sensory input to the central nervous system, and the resulting motor commands to the muscles, should be conveyed with the necessary rapidity. In the non-myelinated fibers of invertebrates, the velocity of conduction is proportional to the square root of diameter, and rapid conduction is achieved simply by an increase in size. My own apprenticeship in the experimental exploitation of giant axons was served almost fifty years ago when I dissected little bundles of half a dozen 30 μm fibers from the walking legs of crabs, and used them for the first measurement of the outward movement of radioactive potassium during the nerve impulse. Giant fibers of this order of size are found in the legs of many crustacea, with somewhat larger ones in their nerve cords, and in the nerve cords of any invertebrate capable of a rapid withdrawal response, such as an earthworm or *Myxicola*.

The prime example is provided by the giant axons up to 1 mm in diameter that radiate from the stellate ganglion to innervate the mantle muscle of the squid, enabling the whole mantle to contract synchronously and expel a jet of water through the funnel beneath the head.

Embryologically, the giant axons of invertebrates are formed by fusing together a large number of smaller axons, and they end up with a cell body of normal size containing a single nucleus in the stellate ganglion, but are surrounded by a sheath about 3 μm thick formed by several layers of loosely packed Schwann cells. They are activated by first-order neurones in the palliovisceral ganglion, connected to second-order giant neurones whose axons form the presynaptic terminals in the stellate ganglion. Because of the large size of the third-order giant axons, a giant synapse is required to excite it, this being another favorite for study by electrophysiologists.

The conduction velocity of a nerve impulse depends essentially on the rate at which a depolarization of the membrane can increase its sodium conductance, and this in turn depends on the membrane capacity and on the density of sodium channels in the membrane. In 1973, what has become known as the sodium gating current was first measured successfully in squid axons by Armstrong and Bezanilla at Woods Hole, and a few months later by Rojas and myself at Plymouth. Subsequent studies have made clear how the gating current arises from the displacement across the membrane under the influence of the electric field of the positively charged arginine and lysine residues carried by the S4 voltage-sensors, and its kinetics have been described in detail by Keynes (1994) and others.

At a Discussion Meeting of the Royal Society held soon after the discovery of the sodium gating current, Sir Alan Hodgkin (1975) pointed out that such a displacement of charge for opening the gates would slightly increase the effective capacity of the membrane, so that there would be a price to be paid as well as a gain for an increase in the density of the sodium channels. He then made some rough calculations of the variation in conduction velocity as a function of channel density and, as seen in Figure 10.2, showed that the optimum channel density was reasonably close to the actual one; his conclusions were later confirmed by Adrian (1975). Hodgkin also pointed out that a similar calculation for the smallest non-myelinated axons like the garfish olfactory

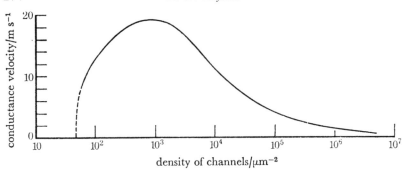

Figure 10.2. Variation of conduction velocity with density of sodium channels calculated by Hodgkin (1975) from the equations of Hodgkin and Huxley (1952). The standard Hodgkin–Huxley axon was assumed to have 500 channels $\mu m^{-2}$.

nerve with a diameter of only $0.2\,\mu m$ would in their case place the channel density a long way from the optimum, and indeed that if the characteristics of such channels were like those of a squid giant axon, the fiber would not actually be able to propagate action potentials at all. However, it is clear that in this case the achievement of maximum conduction velocity does not apply, because the fish always has plenty of time in which to sniff around its surroundings, and that the optimized design of the channels aims at economy of energy consumption with a much slower conduction rate.

## Speeding up nerve conduction by myelinization

Although the evolution of giant axons has evidently proved to be highly advantageous to squid, which are present in countless numbers in the world's oceans, there are many other animals which depend on very rapid motor responses, but which could not afford the space in their bodies for nerve tracts consisting of 1 mm axons. These are, of course, the vertebrates, one of whose greatest evolutionary advantages has been the development of myelinated nerve fibers. Here the passage of local circuit current across the excitable membrane is restricted to the nodes of Ranvier, which are separated by stretches of membrane insulated electrically by the myelin sheath (Figure 10.3). The diameter of the myelinated axon is only 0.05 mm, whereas a non-myelinated axon conducting at the same velocity would be ten times fatter. Since in vertebrates there are also proprioceptor fibers that need to conduct equally

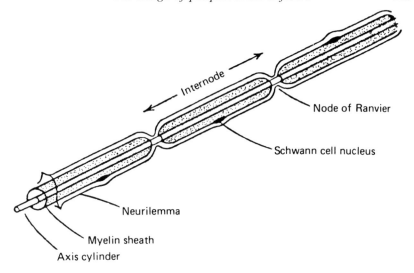

Figure 10.3. Schematic diagram of the structure of a vertebrate myelinated nerve fiber. The distance between neighboring nodes is actually forty times greater relative to the fiber diameter than is shown here. (From Keynes and Aidley 1991.)

fast, a huge saving in bulk has been achieved by the myelinization of peripheral nerve fibers. A few invertebrates, the prawn being the usual example quoted, have crudely myelinated fibers formed by a loose wrapping of Schwann cells around the axon with their cell nucleus on the inside. This does not provide a very effective insulating layer, and the fully developed myelin sheath of vertebrates consists of more than a hundred layers of Schwann cell membrane very tightly wrapped around the axon, with a single cell nucleus on the outside of each internodal stretch of 1.5 mm.

The sodium channels that generate the action potential are confined to the axonal membrane at the nodes of Ranvier, where they are packed even more densely (700 channels $\mu m^{-2}$) than they are in the squid giant axon. This arrangement speeds up saltatory conduction of the nerve impulse by greatly reducing the capacitance per unit length of nerve that has to be charged and discharged as the impulse advances. In the case of myelinated nerve, conduction velocity is directly proportional to diameter, and the calculations of Hodgkin (1975) and Adrian (1975) would not be applicable. I am not aware that the optimum density of sodium channels at the node has been computed by anyone, but I should be surprised if it turned out to be far from 700 channels $\mu m^{-2}$.

## Structural optimization of peripheral nerves

The first and overriding respect in which the structure of peripheral nerves has been optimized lies in digitalization, ensuring that the information conveyed depends on the pattern and number of impulses transmitted by each fiber, and is not at the mercy of a conduction time that might vary from time to time with local conditions. The second respect is that the size or myelinization of the fibers is closely adapted to their specific function, so that the largest ones are preserved for pathways where high speed of conduction is essential for survival, and the smallest and slowest ones are used for sensory pathways where rapidity is not a primary requirement, or for control of the autonomic nervous system.

Some nice examples of specialized symmorphosis are provided by the structures of the electric organs of fishes described by various authors in the volume edited by Chagas and Paes de Carvalho (1961). These have developed from muscle rather than nerve fibers, and in the fresh-water fishes *Electrophorus* and *Malapterurus* the additive discharge of several thousand cells arranged in series depends on voltage-gated sodium channels identical with those shown in Figure 10.1. There are ingenious adaptations to ensure that initiation of the discharge by nerve impulses traveling from a command center in the brain reach all parts of the electric organ at the same moment. In the marine elasmobranch Torpedo, where the electroplates are developed from motor endplates and are not sodium-dependent, the requirement is to generate large currents rather than large voltages, and the electric organ therefore has many columns of cells arranged in parallel instead of large numbers of cells operating in series.

Myelinated nerves can conduct fast enough to meet the needs of motor control in even the largest mammals. It appears that in whales, the ventral horns of the spinal cord are much larger relative to the dorsal horns than in other vertebrates, but this is not thought to reflect a greater size of their motor nerves. It has been suggested that it probably results from the fact that their small surface-to-volume ratio, the absence of limbs, and the constancy of temperature of their environment, means that they have relatively fewer sensory receptors in their skins. However, I am not sure that this argument is valid, any more than the idea that fruit-eating primates need to have larger brains than their leaf-eating relatives because good color vision is needed for the selection of ripe fruit!

## Further reading

Adrian, R. H. (1975) Conduction velocity and gating current in the squid giant axon. *Proc. R. Soc. Lond. B*, **189**, 81–6.

Chagas, C. and Paes De Carvalho, A. (1961) *Bioelectrogenesis. A Comparative Study of its Mechanisms with Particular Emphasis on Electric Fishes.* Elsevier, New York.

Hodgkin, A. L. (1975) The optimum density of sodium channels in an unmyelinated nerve. *Phil. Trans. R. Soc. Lond. B*, **270**, 297–300.

Hodgkin, A. L. and Huxley, A. F. (1952) A quantitative description of membrane current and its application to conduction and excitation in nerve. *J. Physiol., Lond.*, **117**, 500–44.

Keynes, R. D. (1994) The kinetics of voltage-gated ion channels. *Q. Rev. Biophys.*, **27**, 339–434.

Keynes, R. D. and Aidley, D. J. (1991) *Nerve and Muscle*, 2nd edn. Cambridge University Press, Cambridge, UK.

Robertson, J. D. (1960) The molecular structure and contact relationships of cell membranes. *Prog. Biophys.*, **10**, 343–418.

# 10.3 Observing design with compound eyes

## SIMON B. LAUGHLIN

To what extent can we unravel the design of a physiological system? Because the mechanisms responsible for capturing and transmitting pictorial information in the blowfly compound eye have been intensively analysed (Laughlin 1994), this system provides considerable insight into the problem of understanding design. I will consider questions related to symmorphosis that are made apparent in the blowfly compound eye: the use of optimization models to define function and identify constraints, the adaptation of physiological response to changing demands, the matching of constraints and components, the utilization of capacity, energetics, and costs versus benefits. I will close by being brought face to face with two overriding biological factors, fitness and the impediment of inheritance.

### Optimizing the uptake of pictorial information

An eye forms an image of an object, samples this optical image with photoreceptors, and then transmits photoreceptor signals to interneurons. The array of interneurons represents pictorial information as a neural image (Figure 10.4), because each cell signals the quality of light from a small patch of space. In the compound eye of a blowfly this process of neural image formation is straightforward. Following Muller's mosaic theory, a facet on the compound eye represents a pixel, a single spatial element in a picture (Land 1981). The task of building up an achromatic neural image from these pixels is assigned to discrete neural modules, groups of six photoreceptors (R1–6) driving two identified interneurons, the large monopolar cells, L1 and L2 (Figure 10.5). There is one neural module for every facet, so that a large monopolar cell's electrical signal represents the gray level in one pixel. In a

# neural image   neural image

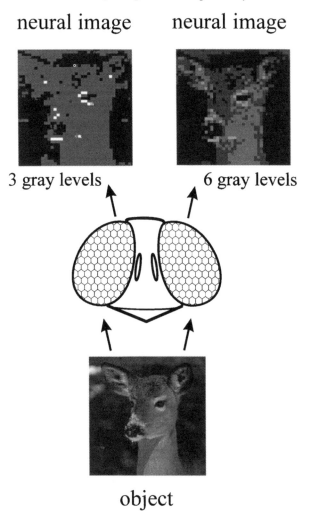

object

Figure 10.4. The quality of the neural image sent from eye to brain increases with the range of signals transmitted by neurons. The fly's compound eye samples an object relatively coarsely so that images are represented as mosaics of discrete samples. Each sample is equivalent to a pixel in a computer-generated image, and corresponds to a single facet on the surface of the eye. Beneath each facet, a small set of six photoreceptors (R1–6) and two interneurons, the large monopolar cells L1 and L2 (Figure 10.5), generate an electrical signal that codes the gray level of their particular pixel. Increasing the strength and reliability of this neural signal increases the range of gray levels in the neural image generated by the array of large monopolar cells. As illustrated by the top two images, this expansion of gray level improves the neural image, and increases the quantity of pictorial information transmitted from eye to brain.

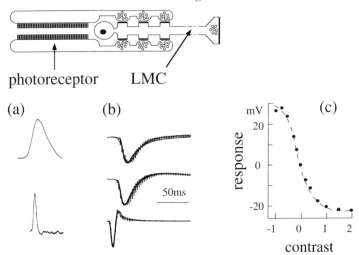

Figure 10.5.   The coding of pixel gray level and its optimization. Pixel gray level is coded by a small neural module: six photoreceptors making synapses with two output neurons, the large monopolar cells L1 and L2. For clarity, only two photoreceptors and one monopolar cell are depicted in this diagram. The six photoreceptors receive light from the same small patch of the object, corresponding to a single facet of the compound eye. The number of gray levels coded by the module depends upon the quality of the electrical signals generated and transmitted by this module. To maximize gray level range, the performance of photoreceptors, synapses, and large monopolar cells are optimized. The analog signals used in this system are illustrated by the intracellular responses of photoreceptors (a) and large monopolar cells (LMC, b) to brief (< 2.5 ms) flashes delivered at the beginning of each trace, either in darkness (top traces), in the presence of sustained full daylight (bottom traces), or for the middle trace in (b), in dim light. All flash responses are normalized to unit peak amplitude. Note that adaptation to daylight accelerates the photoreceptor response, leading to a brisker response in the large monopolar cell, and adds a depolarizing component to the monopolar cell waveform. The means and standard deviations of responses to many repetitions of identical stimuli (vertical lines) fit the continuous curves which define the response of the optimal filter predicted by information theory (van Hateren 1992). (c) The relationship between the amplitude of response and stimulus contrast for the light-adapted system. The measured responses (●) fit the cumulative probability distribution for contrast in natural scenes (- - -) almost exactly, so adhering to the optimum predicted by information theory (Laughlin 1981). (Data in (b) replotted from van Hateren, J. H. (1992) *Nature*, **360**, 68–9.)

given eye the number of facets, and hence pixels and neural modules, is fixed. Consequently, to increase the pictorial information in the neural image, the range of pixel gray-tones signaled by the large monopolar cells must be increased (Figure 10.4). Increasing the speed with which signals change also boosts the rate of information uptake. Similar goals of maximizing pixel signal range and refresh rate challenge the designers of visual display units.

To understand how the formation of the large monopolar cells' neural image is optimized, we must first consider the signals that represent gray-tone. Photoreceptors collect light from the facet and convert photon absorption rate into membrane potential, using a biochemical amplifier. This phototransduction cascade couples photon capture by a rhodopsin molecule to the opening of Na/Ca channels in the membrane, and adapts its gain to changing load. In dim light one absorbed photon opens several hundred channels to produce a detectable response but this ratio drops in bright light, to prevent saturation. The six photoreceptors send their signals to a large monopolar cell across an array of 1,320 chemical synapses, 220 per photoreceptor. Synaptic transmission reverses signal polarity but, because action potentials are not involved, this inversion has no effect on information content.

Photoreceptors and large monopolar cells (LMCs) are using analog signals, continuous modulations in membrane potential, to represent the depth and time course of changes in gray-tone. The compound eye regulates the amplitude, reliability, and waveform of responses, to optimize the rate at which LMCs transmit pictorial information to the brain. This optimization is performed within a demanding set of neural constraints. Ionic mechanisms limit the range of analog signals to 60 mV. Discrimination of signal levels within this range is limited by intrinsic noise; for example, random fluctuations generated at the synapses. Moreover, the system adjusts to seven orders of magnitude of input, from a behavioral threshold at 4 photons receptors$^{-1}$ s$^{-1}$ to full daylight.

Is signaling optimized? Given inputs of a certain size and shape, as defined by the statistics of naturally formed optical images, information theory gives the number of different signals coded within the LMC's response range, and identifies optimum coding strategies that maximize information uptake (Laughlin 1981). For example, the curve relating response amplitude to input level follows the cumulative probability distribution of input amplitudes (Figure 10.5c), and this precise relationship optimizes the utilization of response range.

Efficiency is increased when one eliminates redundant signal components that carry little or no information (for example, levels that remain constant over large intervals of space or time). Given the statistical "shapes" of inputs, as determined by the distribution of natural objects, locomotory patterns, and photoreceptor quantum catch, information theory predicts an optimum filter that rejects redundant components and maximizes the information packed into LMCs (van Hateren 1992). The operation of this filter is quantified by its impulse response, the waveform generated by a small instantaneous light-flash. In dim light this optimum filter responds slowly, to smooth out photon noise (the unwanted fluctuations introduced by the Poisson process of photon absorption). At high light levels the photon flux is more reliable. Consequently, the optimum filter retunes to a brisk differentiator that emphasizes change and rejects constant components. LMCs conform to this theoretical optimum with remarkable fidelity (van Hateren 1992), retuning their responses to suit the ambient light level (Figure 10.5b). In conclusion, two remarkably close fits to optimization models (Figure 10.5b,c) have suggested a function for photoreceptors and LMCs (the efficient coding of pictorial information), identified constraints (noise, limited neural response range), and demonstrated the importance of adaptation to changing conditions. Indeed, it is this adaptation of response (Figure 10.5b) that provides the most stringent test of optimum coding.

### Matching components and constraints

Three cellular mechanisms – phototransduction cascade, photoreceptor membrane, and photoreceptor synapse – are co-regulated to optimally tune LMCs. At low light levels the phototransduction cascade responds sluggishly to produce the slow response that optimally smoothes noisy signals. Light adaptation accelerates the cascade, reducing the response duration from more than 50 ms to approximately 10 ms, to produce the crisp response that is optimum in bright light (Figure 10.5a). To accommodate the rapid response of the light-adapted cascade, the photoreceptor membrane is tuned by voltage-gated potassium channels. These channels open when the photoreceptor is depolarized by light to reduce membrane resistance, and hence the membrane time constant. Exposure to bright light also facilitates adaptation at the synapse, leading to a biphasic response in the LMC. Thus, by co-regulating these separate mechanisms, the overall performance of the LMC is adapted to operating conditions to optimize performance.

The voltage-gated potassium channels are matching a physical constraint, the membrane time constant, to the quality of the signal that drives the membrane. Indeed, the potassium channels have precisely the combination of properties required to match the gain and response speed of the membrane to the phototransduction cascade's: a high gain and slow response in the dark; a low gain and fast response when depolarized by light (Weckström and Laughlin 1995). These potassium channels appear to have been selected for this regulatory role. Comparative studies, a useful tool for probing design, show that slowly flying Diptera have photoreceptors that fail to speed up with light-adaptation. In the absence of fast-moving signals, this slow response is better and the photoreceptors are using inactivating potassium channels to save energy, even in bright light (Weckström and Laughlin 1995).

## The utilization of capacity

The ability of photoreceptors to transmit information is limited by the amount of molecular machinery, much as sugar uptake in the gut is limited by the number of membrane-bound transport molecules. For a fly in daylight, photoreceptor reliability reaches a ceiling set by a finite population of about 50,000 transduction units per cell. Unnaturally bright light saturates this finite population of signal-generating units, leading to a dramatic decline in signal (Howard, Blakeslee and Laughlin 1987). To prevent saturation, an optical feedback mechanism, analogous to our own pupil, attenuates light within photoreceptors. This pupil maintains the photon absorption rate at a level that makes optimum use of the limited number of transduction units over a 100-fold range of daylight levels. Following a principle of symmorphosis, insect compound eyes demonstrate that structures are enlarged to increase capacity. The 1 mm long photoreceptors of the dragonfly are four times longer than the blowfly's to provide more transduction units, and hence a better SNR (Laughlin 1994).

Just like transduction units, the number of synaptic vesicles carrying the signal from photoreceptor to large monopolar cell limits information capacity. The random process of vesicle release introduces unreliability, but this synaptic noise is kept within reasonable bounds by using a massive synaptic array, 220 per photoreceptor (Figure 10.5), to provide hundreds of thousands of vesicles per second for signal transfer. In addition, the optimum filtering of signals, described above, ensures that this population of vesicles is used to its best advantage by eliminating wasteful redundancy (Laughlin 1994). Again the available machinery is limiting and the system is designed for full utilization. At high light levels, synaptic noise approaches levels set by transduction units in photoreceptors. Are these two separate constraints matched by structural adjustments to the number of synapses? A glance at the literature suggests that the longer photoreceptors with more transduction units make more synapses, but this observation should be quantified.

### Costs, eye design, and limits to optimality

In the blowfly's compound eye one can measure the metabolic cost of sensory information (Laughlin, de Ruyter van Steveninck, and Anderson, submitted, *Nature*). Costs are high. The excellent performance of blowfly photoreceptors depends on a large light-gated conductance, driven by 50,000 transduction units, and a counterbalancing potassium conductance. These high conductances generate large ionic fluxes that are restored by pumps which consume 10 percent of the total oxygen consumption of a resting fly (Howard *et al.* 1987). In flight the fly's oxygen consumption rises 30-fold, and the relative cost of transduction becomes negligible. However, from the estimated retinal mass, 2 percent of total oxygen consumption is used to carry the eye in flight. No wonder the eye's optical and neural components are optimized, and eye size varies from species to species.

To what extent can these high metabolic costs of vision explain interspecific differences in eye size and optical design? Taking the cost of the eye to be proportional to volume, as suggested for the flying fly, one can use simple scaling arguments (Kirschfeld 1976), based upon optical limitations to visual acuity, to determine the cost of information (Figure 10.6). The cost of the fundamental pictorial building block, a pixel, rises dramatically with the total number of pixels acquired. Because the law of diminishing returns holds sway, optics alone cannot account for investment in vision. The ultimate arbiter of eye size must be the biological value of that information. This conclusion draws attention to an important distinction, recognized in economics, between the currency of information theory (bits, pixels) and the utility of information. To give a simple example, information theory defines the number of bits required to send the Dow-Jones Index from New York to London, but will not tell you how to make a killing on the market.

The cost of information illustrates another general limitation to optimization arguments in biology: the range of eye designs available to an insect is constrained by inheritance. Compound eyes are much less efficient than single lens eyes (Kirschfeld 1976). Bigger lenses catch more light and have a better resolving power, giving a much lower cost per pixel (Figure 10.6). Only a process that is ignorant of the laws of optics would select an array of small lenses (an apposition compound eye) in

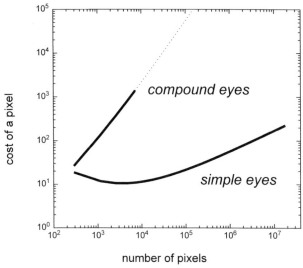

Figure 10.6. The unit cost of vision, the cost of a pixel, plotted as a function of the number of pixels (that is, the spatial sampling density) for apposition compound eyes and lens eyes. The dotted line extrapolates the cost of pixels beyond the dimensions of the largest extant compound eyes. The cost is taken to be the volume of the eye corresponding to one pixel, given in units of $1000\,\mu m^3$. The number of pixels is calculated for an array viewing one steradian of solid angle. For both eye types the F-ratio is fixed at 6, and the photoreceptors have a diameter $2.5\,\mu m$ and length $100\,\mu m$. Sampling is at the diffraction limit but undersampling, a common strategy in compound eyes (Land 1981; Laughlin 1994), simply shifts the curves slightly to the left. Photon noise is assumed to be negligible. The volume of the compound eye is calculated by assuming that it is a sheet, with a surface area equal to that of the array of square facets, and a depth equal to the focal length of the lens plus the photoreceptor length. Tapering the ommatitida will shift the curve down slightly. The simple eye is a hemisphere with a radius equal to the lens focal length plus the photoreceptor length ($100\,\mu m$). The addition of receptor length to the dimensions of the eye produces a minimum unit cost at a lens diameter $\approx 30\,\mu m$. Shortening the receptors reduces the optimum radius of the simple eye.

preference to a single large lens (a simple eye). Given that almost every other feature of a compound eye is superbly designed to utilize resources and minimize constraints, why are insects using this inferior optical instrument?

The constraint of inheritance provides a plausible explanation. Starting with an eye with one pixel, duplication increases sensitivity and, following adjustments to data processing, allows for simultaneous spatial discriminations. These are great advantages, irrespective of whether they are achieved by duplicating the entire photodetector, receptor and optics, to make a compound eye, or duplicating photoreceptors beneath the original lens, to make a simple eye. As pixels multiply, the brain evolves to handle the richer images and animals see better. By the time the inferior design of the compound eye begins to tell, throwing away the brain and starting again is not an option. The argument that many arthropods are locked into an inferior design by their brain is strengthened by observing that insects also have simple eyes with good large lenses and well-developed retinas. Despite their optical potential, these ocelli are defocused for detecting the horizon, and their neural projections lack the columnar, pixel-mapping structure of the optic ganglia.

### Conclusion: constraints upon our understanding of design

Theory and experiment demonstrate that, among compound eyes, the blowfly's is superbly designed. Performance is optimized within proximate constraints, namely the physics of natural images and the biophysics of neural communication. These workings illustrate many elements of symmorphosis: adaptation to changing load, matching components and constraints, and the utilization of capacity. However, elementary considerations of optics and eye mass demonstrate that the unit cost of information cannot, in isolation, define how much an animal should invest in vision. Although obvious advantages (for example, large eyes for night vision and for detecting prospective prey or mates) can be quantified in terms of sensitivity and acuity (Land 1981), the value of good vision must ultimately be related to fitness. Furthermore, the inferior optical design of compound eyes suggests that developmental and operational constraints have obstructed the evolution of superior designs. We see, through a fly's eye, that physiology is well equipped to demonstrate that systems are optimized within the proximate constraints of mechanism, but the simplest attempts to broaden this quantitative approach founder on two ultimate constraints, fitness and phylogeny.

### Further reading

Howard, J., Blakeslee, B. and Laughlin, S. B. (1987) The intracellular pupil mechanism and photoreceptor signal-to-noise ratios in the fly *Lucilia cuprina*. *Proceedings of the Royal Society of London B*, **231**, 415–35.
Kirschfeld, K. (1976) The resolution of lens and compound eyes. In *Neural Principles in Vision*, pp. 354–70. Eds. F. Zettler and R. Weiler. Springer-Verlag, Berlin, Heidelberg and New York.

Land, M. F. (1981) Optics and vision in invertebrates. In *Handbook of Sensory Physiology*, vol. VII/6B, pp. 471–592. Ed. H.-J. Autrum. Springer-Verlag, Berlin, Heidelberg and New York.

Laughlin, S. B. (1981) A simple coding procedure enhances a neuron's information capacity. *Zeitschrift für Naturforschung*, **36c**, 910–12.

Laughlin, S. B. (1994) Matching coding, circuits, cells, and molecules to signals – general principles of retinal design in the fly's eye. *Progress in Retinal and Eye Research*, **13**, 165–96.

Van Hateren, J. H. (1992) Theoretical predictions of spatiotemporal receptive-fields of fly LMCs, and experimental validation. *Journal of Comparative Physiology,* **A171**, 157–70.

Weckström, M. and Laughlin, S. B. (1995) Visual ecology and voltage-gated ion channels in insect photoreceptors. *Trends in Neurosciences*, **18**, 17–21.

## 10.4 Evolution of a visual system for life without light: optimization via tinkering in blind mole rats

EVIATAR NEVO

Does natural selection globally optimize structure and function; that is, does it work as an engineer or does it operate adaptively via tinkering, producing local optimization? As Darwin repeatedly demonstrated in the *Origin of Species*, the living world contains many structural and functional imperfections. Evolution appears not to have reached perfection. In contrast to the engineer, who assembles raw materials and tools to fit his project, the tinkerer always works with the available material, constraints, and contingencies. Evolution, as emphasized by Jacob (1977), behaves like a tinkerer, slowly modifying adaptively available molecular and organismal structures to generate emergent novelties.

### Eye evolution and regressive adaptations

Subterranean mammals provide a global natural experiment of life without light. The family Spalacidae of southwest Asia, which originated in Oligocene times around 30 million years ago and to which the genus *Spalax* belongs, provides an example of extreme adaptation to life underground which went along with a severe regression of the eyes and the visual system.

Complete disappearance of the eyes has not occurred in the evolution of any mammalian species. Maintenance of eye rudiments after millions of years of adaptation to a lightless environment might therefore suggest that such rudiments still have some functional role in the development of the animal.

The rudimentary eyes of the mole rat, belonging to the genus *Spalax*, are subcutaneous, as the eyelids are permanently sealed (Figure 10.7), and do not respond to light stimuli. They demonstrate no visual cortical potentials and show no overt behavioral response to light. However, removal of

Figure 10.7.   One of the four chromosomal, morphologically indistinguishable, species of mole rats, *Spalax ehrenbergi* superspecies in Israel. Note the sealed eyelids.

the eyes disturbs photoperiod perception in these animals. Mole rats have a circadian rhythm entrainable by light. The eye participates in biorhythmicity, and in thermoregulation. Thus, the *Spalax* eye provides a remarkable evolutionary model to study reduction and loss of function (vision), emphasizing its remaining function (photoperiodic perception) as an organ.

In the early embryos the presumptive eye regions – the epithelium, lens vesicle, and optic cup – appear initially normal (Figure 10.8). As development progresses, the iris–ciliary body complex originates prematurely from the margin of the optic cup and shows a very rapid and hypertrophic growth. In contrast, retinal histogenesis progresses relatively normally, and results in structurally reduced but well-differentiated photoreceptor, neuronal, and ganglion cell layers in the adult eye (Figure 10.8). Immunohistochemically, the presence of opsin was demonstrated in the photoreceptor cells. The normal retina and opsin may indicate that these rudimentary eyes are still functioning in the complex neuroendocrine pathways mediating photoperiodicity, primarily during transition from short to long photoperiod.

### Ocular regression conceals adaptive progression of the visual system of *Spalax*

The visual system of *Spalax* presents a mosaic of both regressive and progressive morphological features. Reduction of the visual system is

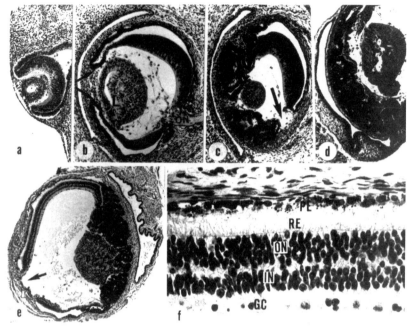

Figure 10.8.    Light micrographs showing cross-sections of the developing eye of
the mole rat, *Spalax ehrenbergi*. (a) Optic cup and lens vesicle initially develop
normally (original magnification ×100). (b) Eye at a later embryonic stage. Note
the appearance of iris–ciliary body rudiment (arrows), and development of the
lens nucleus (L). ON, optic nerve (original magnification ×100). (c) Eye at a still
later fetal stage. Note the massive growth of the iris–ciliary body complex and
persistent colobomatous opening (arrows) (original magnification ×100). (d)
Early postnatal stage. The iris–ciliary body complex completely fills the anterior
chamber. The lens is vascularized and vacuolated (original magnification ×100).
(e) Adult eye. Eyelids are completely closed, and a pupil is absent. Note the
atrophic appearance of the optic disc region (arrow) (original magnification
(×65). (f) Higher magnification of the adult retina. The different retinal layers
are retained: PE, pigment epithelium; RE, receptor layer; ON, outer nuclear layer;
IN, inner nuclear layer; GC, ganglion cell layer (original magnification ×500).
(Reproduced from Sanyal, S., Janse, H. G., de Grip, W. J., Nevo, E. and de Jong,
W. W. (1990) The eye of the blind mole rat, *Spalax ehrenbergi*: rudiment with
hidden function? *Invest. Ophthalmol. Vis. Sci.*, **312**, 1398–404. © Association for
Research in Vision and Ophthalmology.)

selective and operates at the level of specific neuronal networks. Thalamic
brain structures of the "image-forming" visual pathways are drastically
reduced, whereas hypothalamic brain structures associated with photoperiodic perception are expanded. The retina, which involves only 823
ganglion cells (compared to 65,000 to 100,000 in other rodents), was

found to project bilaterally to all visual structures described as receiving retinal afferents in non-fossorial rodents. Brain structures involved in "image forming" and visually guided behaviors are reduced in size by more than 90 percent, receive sparse retinal innervation, and are cytoarchitecturally poorly differentiated.

The eye is not globally reduced. In sharp contrast to the foregoing regression of the "image-forming" visual pathway, brain structures of the "non-image-forming" visual pathway, involved in photoperiodic perception, are well developed in *Spalax* (Figure 10.9). The suprachiasmatic nucleus (SCN), the circadian pacemaker in the mammalian brain, receives a bilateral projection from the retina, and the absolute size, cytoarchitecture, density, and distribution of retinal afferents in *Spalax* are comparable with those of other rodents. These results indicate that the apparently global morphological regression of the visual system conceals a selective expansion of structures related to functions of photoperiodic perception and photo-neuroendocrine regulation.

We argued that far from sheer degeneration, the evolution of an atrophied eye and reduced visual system of *Spalax* is an adaptively advantageous response to the unique subterranean environment. Factors favoring regression include mechanical aspects, metabolic constraints, and competition between sensory systems. The primary advantage of sensory atrophy is the metabolic economy gained by the reduction of visual structures that do not contribute significantly to the animal's fitness. Eye metabolism (mainly the photoreceptor cells) is 100 times that of other body tissues. While the eye is a fraction of body weight, this is still a very large value. In addition, the eye is not an isolated organ – it is directly connected to structures within the brain which are also highly metabolically active. Reduction of the eye not only reduces the retina; the corresponding structures related to vision are also reduced.

## Brain evolution of *Spalax*

To date, no visual cortex has been identified, either structurally or electrophysiologically. Instead, the occipital cortex of *Spalax* is occupied by an extension of the somatosensory cortex. Likewise, the somatosensory thalamus of *Spalax* almost reaches the dorsolateral surface, without evidence of the lateral geniculate body, the first station in the visual pathway. These findings indicate that the thalamocortical visual system in *Spalax* is minute, whereas the somatosensory system is expanded. Clearly, the brain area normally occupied by the visual cortex serves

292    *E. Nevo*

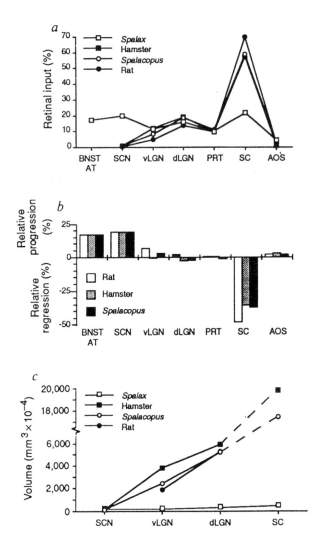

Figure 10.9. (a) The relative degree of retinal input to primary visual structures compared to the total quantity of retinal projections in *Spalax*, hamster, rat, and *Spalacopus cyanus* (South American Octodontidae, "cururo"). All these rodents are of similar body size (120–140 g). The density of retinal projections in each primary visual structure was quantified using a previously described method. (b) The relative degree of change in the proportions of retinal input to different primary visual structures in *Spalax* compared to measures obtained in the other rodents. A relative progressive development in *Spalax* is seen in structures involved in photoperiodic functions (SCN, BNST). The main regressive feature is the drastic reduction of retinal input to the superior colliculus. The relative size of

somatosensory functions. These, in addition to the prevalence of seismic communication in *Spalax* evolution, suggest that touch replaced vision as a major communication modality in the underground life of *Spalax* and probably in other subterranean mammals.

## Summary of evidence

*Spalax* is an extreme example of natural visual degeneration in mammals (Figure 10.9). The visual cortex has been replaced by the somatosensory cortex. Visual pathways are regressed, and the absence of visual cortical potentials and overt behavioral response to light indicate that *Spalax* is blind. Structural, molecular, and experimental evidence indicate a functional role for the retina in light perception. Entrainment of circadian locomotor and thermoregulatory rhythms by ambient light indicates the capacity for photoperiodic perception. The atrophied subcutaneous eyes underwent mosaic adaptive evolution: severe regression of thalamic and tectal structures involved in form and motion perception is coupled to hypertrophy of hypothalamic structures subserving photoperiodic functions. Ganglion cell retinal projections to the SCN are increased twentyfold, and the SCN, in which the circadian pacemaker resides, is identical in *Spalax* and sighted rodents.

## Stochasticity and selection in evolution

The problem of vestigial organs is an old one, discussed by Darwin and elaborated more recently. Is eye evolution in *Spalax*, and other microphthalmic organisms, determined by stochastic or deterministic evolutionary factors? Or both? Or, more generally, are structural reductions of "useless vestigial organs" in evolution determined largely by random mutation pressure, or rather channeled by directional selection that increases fitness?

other visual structures in *Spalax* are unmodified compared to that of the other species. (c) Comparison of the absolute size (volume, mm$^3$ × 10$^{-4}$) of several visual structures in *Spalax* and other rodents. The size of the SCN is equivalent in all species. The vLGN and dLGN are reduced by roughly 90 and 94 percent, respectively. The retino-recipient layer of the superior colliculus is reduced by 97 percent. Values for the rat are adapted from Sugita and Otani (1983). (Reprinted with permission from *Nature*, **361**, Cooper, H. M., Herbin, M. and Nevo, E. (1993) Ocular regression conceals adaptive progression of the visual system in a blind subterranean mammal. © 1993 Macmillan Magazines Ltd.)

I will argue that in the dramatic case of mosaic eye evolution in *Spalax*, and other subterranean mammals, and probably more generally in evolution, structural reductions and expansions are adaptive and are channeled by directional selection for reduced but functional eyes, at the level of specific neuronal networks.

## Selection pressures for reduced (microphthalmic) eyes

Eye regression in invertebrates and vertebrates is generally related to "life in a dark environment," although the adaptive significance is obscure. Hypotheses explaining regression involve either environmental or genetic factors. The former explain eye regression as a direct consequence of lack of light on the eye during development. Consistent with this explanation is the significant selection for reduced eyes demonstrated in cave populations. Inconsistent, however, is the finding that eye regression occurs when microphthalmic species are raised in constant light.

## Mechanical, metabolic, and competition determinants of *Spalax* eye evolution

Clearly, large exposed eyes are an impediment to a subterranean mammal. Selection for microphthalmy would minimize infection or injury. Diverse morphological adaptations occur in subterranean mammals to protect the eye. Likewise, vision is useless in the dark underground world where subterranean mammals are sealed most of their lives (more than 99 percent of the lifetime of *Spalax*). Strong negative selection operates to avoid the useless maintenance of an obsolete neural network.

Despite its relatively small size, the metabolic expense of the brain is high: it consumes more than 20 percent of $O_2$ and glucose. Visual structures (dLGN, superior colliculus, visual cortex) exhibit the highest rates of local glucose metabolism in the brain. Reduction in *Spalax* of visual structures substantially economizes metabolic costs. These structures are reduced in *Spalax* by 90–99 percent. In other words, bioenergetics and metabolic economy are critical, particularly in the low productive and hypoxic underground environment. Superfluous neurons are a luxury strongly selected against and their reduction (or replacement) can contribute to fitness by improving metabolic efficiency.

The dramatically reduced retina (to 1.33 mm$^2$ (99 percent), with its 823 ganglion cells) in *Spalax* is highly specialized. Specifically adapted are the high molecular opsin in *Spalax* photoreceptors and the mid- to long-wave sensitive cone opsin involved in photoentrainment, which contribute to light detection. Notably, removal of the eye impedes thermoregulation and photoentrainment. *Spalax* circadian rhythmicity is presumably genetically determined by *per* involvement through photic induction of *Fos* immunoreactivity in the suprachiasmatic nucleus where the "zeitgeber," or pace-maker of circadian rhythmicity, resides. The hormone melatonin, generated in the retina, pineal, and harderian glands, was shown to be involved in photoperiodic perception, improvement of thermoregulation, and biorhythmicity.

Regressive evolution of the sight in *Spalax* was complemented by the progressive evolution of nonvisual sensory systems for communication, foraging, and orientation. These nonvisual systems include magnetic orientation, vocal, seismic (vibratory), olfactory, and sensual modalities. Both call and seismic (vibrational) signals are optimally transmitted underground through the evolution of low-frequency signals. Olfaction is acute, and magnetic orientation contributes to orientation in underground intricate tunnels. These progressive physiological functions are supported by structural progression and reorganization of the middle ear ossicles and the cochlear receptors, the vestibular organ, as well as brain motor and somatosensory structures. These progressive evolutionary processes further favored reduction of the visual system.

## The evolutionary conservation of primary visual organization of the *Spalax* eye

Despite dramatic reduction of "image forming" structures and expansion of "non-image forming" photoreceptor structures, the basic pattern of primary visual organization in *Spalax* is conserved and complete. This may derive from developmental interdependence requiring sequential appearance of individual components for appropriate eye, lens, or brain development or from pleiotropic gene effects. Importantly, mutations can result in phenotypic variation, or change in nucleotide replacement rate as in aA-crystallin, but cannot result in selective deletion of the visual system without leading to a profound reorganization and lethality.

### Mutant strains of anophthalmic rodents

Sporadic and hereditary anophthalmia and microphthalmia resulting in developmental degenerative eye modification have been reported, such as in "eyeless" mice. In the latter, a developmental inhibition of the optic vesicle at around 10 days of age results in complete eye disappearance after 13 days. Retinal ganglion cells never develop and optic fibers never attain the brain. In *Spalax*, however, embryonic eye development appears normal up to 10 days, but this is followed by a number of degenerative processes occurring between 10 and 13 days of embryonic age. Nevertheless, retinal histogenesis is normal, and results in reduced, but well-differentiated, cell layers. The comparable embryonic onset of degenerative processes of the eye in *Spalax*, anophthalmic, and microphthalmic mice suggests the involvement of analogous or homologous embryonic genetic mutations.

## The genetics of loss-and-gain function mutations

A *Drosophila* gene that contains both a paired box and a homeobox with extensive sequence homology to the mouse *Pax-6* (*Small eye*) gene was recently isolated and mapped to chromosome IV in a region close to the eyeless (*ey*) locus. The finding that *eyeless* of *Drosophila*, *Small eye* of the mouse, and human *Aniridia* are encoded by homologous genes suggests that eye morphogenesis is under similar genetic control in both vertebrates and insects, in spite of the large differences in eye morphology and mode of development.

These results support the proposition that *ey* is the master control gene for eye morphogenesis. Because homologous genes are present in vertebrates, ascidians, insects, cephalopods, and nemerteans, we can conclude that *ey* functions as a master control gene throughout the metazoa. All morphogenetic eye genes appear to be under the direct or indirect control of *ey*, which is at the top of the regulatory cascade or hierarchy. A parallel situation in rodents applies to the *Sey* = *Pax-6* gene which is expressed first and controls a set of subordinate regulatory genes, including genes that influence cell–cell interactions and signal transduction. The *Sey* gene in the mouse may be the master control gene in the mouse eye induction process, and hence also in *Spalax*.

The observation that mammals and insects, which have evolved separately for more than 500 million years, share the same master control gene for eye morphogenesis indicates that the genetic control mechanisms of development are much more universal and conserved than anticipated.

The most fundamental distinction between *Spalax* and mutant or enucleated experimental animals is that regression in *Spalax* is uniquely focused on the "image-forming" system; the component of the photoperiodic, "non-image-forming" system in *Spalax* remains intact. The differential genetics underlying regression and progression in the *Spalax* eye is unknown. However, it is likely to be explained by the combinatorial interaction of several homeotic genes. Mutations in some of these genes may lead to a reduction of eye and brain structures controlling image formation. Other homeotic genes may affect expanded eye and brain structures controlling photoperiodic perception. Both gene groups may work epistatically and be under the *Sey = Pax-6* master control gene involved in eye morphogenesis. Future research will decipher the underlying genetic basis of mosaic eye evolution in *Spalax*, on the basis of mutant phenotypes in the *Hox* and other homeotic gene clusters.

### Conclusions and prospects: optimization theory and eye evolution

Optimization models help us to test our insight into the biological constraints that influence the outcome of evolution. They serve to improve our understanding about adaptations, rather than demonstrate that natural selection produces optimal solutions. Two alternative approaches, quantitative genetics, which helps to identify constraints needed for a satisfactory optimization models, and the comparative method, assist in testing predictions of optimality models. Optimality theory has been criticized by theoretical geneticists focusing on the genetic problems of reaching optima. Adaptationists, on the other hand, use optimization models as a tool to understand adaptation, for example, by comparing predictions of game theoretic and diploid genetic models.

Molecular biology dramatized and revolutionized the evolutionary studies of the eye. Loss-of-function mutations in both the insect and in the mammalian genes have been shown to lead to a reduction or absence of eye structures, which suggest the function of *ey* as a master control gene for eye morphogenesis across the metazoa. Could it be that *ey* is the master control gene for both reduction and expansion hypertrophy in the mosaic eye evolution of *Spalax*?

Clearly, the evolutionary tinkering processes of regression and progression of subterranean mammals in general, and particularly of their complex mosaic eye evolution, climaxing in *Spalax*, are now wide open to molecular-genetic experimental studies. While local optimization may be the evolutionary "goal," certainly molecular-genetic tinkering within phylogenetic constraints underlies mosaic eye evolution, as it does in the evolutionary process in general.

## Acknowledgments

I am very grateful to all my collaborators in the long-term project of *Spalax* eye and brain evolution. Without their high-quality expertise this project could never have been realized.

I thank the Human Frontier Science Program grant (GR-68/95 B Cooper), the Israeli Discount Bank Chair of Evolutionary Biology, and the Ancell-Teicher Research Foundation for Genetics and Molecular Evolution, for their financial support of this research.

## Further reading

Argamaso, S. M., Froehlich, A. C., McCall, M. A., Nevo, E., Provencio, I. and Foster, R. G. (1995) Photopigments and circadian system of vertebrates. *Biophys. Chem.*, **56**, 3–11.

Cooper, H. M., Herbin, M. and Nevo, E. (1993) Visual system of a naturally microphthalmic mammal: the blind mole rat, *Spalax ehrenbergi. J. Comp. Neurol.*, **328**, 313–50.

De Jong, W. W., Hendriks, W., Sanyal, S. and Nevo, E. (1990) The eye of the blind mole rat (*Spalax ehrenbergi*): Regressive evolution at the molecular level. In *Evolution of Subterranean Mammals at the Organismal and Molecular Levels*, pp. 383–95. Eds. E. Nevo and A.O. Reig. Alan R. Liss, New York.

Dupré, J. (Ed.) (1987) *The Latest on the Best. Essays on Evolution and Optimality*. MIT Press, Cambridge, MA.

Feldman, J. F. (1988) Genetics of circadian clocks. *Bot. Act.*, **101**, 128–32.

Gehring, W. J., Affolter, M. and Burglin, T. (1994) Homeodomain proteins. *Annu. Rev. Biochem.*, **63**, 487–526.

Jacob, F. (1977) Evolution and tinkering. *Science*, **196**, 1161–6.

Nevo, E. (1991) Evolutionary theory and processes of active speciation and adaptive radiation in subterranean mole rats, *Spalax ehrenbergi* super-species in Israel. *Evol. Biol.*, **25**, 1–125.

Pevet, P., Heth, G., Haim, A. and Nevo, E. (1984) Photoperiod perception in the blind mole rat (*Spalax ehrenbergi*, Nehring): involvement of the harderian gland, atrophied eyes and melatonin. *J. Exp. Zool.*, **232**, 41–50.

Rehkamper, G., Necker, R. and Nevo, E. (1994) Functional anatomy of the thalamus in the blind mole rat *Spalax ehrenbergi*: an architectonic and electrophysiologically controlled tracing study. *J. Comp. Neurol.*, **347**, 570–84.

Sanyal, S., Jansen, H. G., de Grip, W. J., Nevo, E. and de Jong, W. W. (1990) The eye of the blind mole rat, *Spalax ehrenbergi*: rudiment with hidden function? *Invest. Ophtalmol. Vis. Sci.*, **312**, 1398–404.

Sugita, S. and Otani, K. (1983) Quantitative analysis of the lateral geniculate nucleus in the mutant microphthalmic rat. *Exp. Neurol.*, **82**, 413–23.

# 11

## How good is best? Some afterthoughts on symmorphosis and optimization

### EWALD R. WEIBEL

The main controversial issue of this book is whether symmorphosis as a theory of optimized design is a useful concept for the understanding of the relations between design and function, in other words, to test whether and to what extent the design tends to be qualitatively and quantitatively optimal for the required function. This notion is taken by some as a premise for their investigations to see how far it is true – and it is rejected even as a possibility by others. If we look at the sociology of science we find that the first group constitutes physiologists who try to understand how the organism and its parts work, whereas the second group comprises evolutionary biologists who explore the basis of design as a consequence of natural selection. The physiologists find that the organism does its jobs remarkably well and wish to understand the mechanisms of functional processes and their control, for example by design properties, whereas evolutionary biologists have become convinced that evolution by natural selection cannot lead to optimal solutions beyond some chance events. Here is a cause for conflict which needs to be resolved if we are not to fall into the trap of a fight of faiths about adaptation, or of the "spandrels of San Marco" where we do not respect the hierarchy of constraints upon the system (Gould and Lewontin 1979).

### Two types of constraints

There is probably agreement even among evolutionary biologists that evolution by natural selection in the sense of Darwin is basically an optimization process on a grand scale, but it is also clear that optimality in the sense of "perfect adaptation" of all traits is hardly ever achievable because of the many opposing influences, both inside and outside the body. Particularly the external, environmental pressures or resistances are unpredictably variable in time and space, so that some trait that now appears far from optimum may have been well adapted in some distant past, and vice versa.

Does the conclusion that optimal adaptation is unlikely mean that we should not even begin to ask the question whether optimization of design is a principle in adjusting the conditions for functional performance and how it works? I do not think so, mainly because this point of view is too narrow. It disregards the possibility that even though an optimal state is not achieved, optimization of design may still be at play. It also disregards the point that the organism is not only faced with *external constraints* such as environmental pressures or resistances, but also with *internal constraints*, which we could perhaps call *constitutional* constraints. It is particularly noteworthy that the internal constraints are due to the highly conservative internal constitution of the body. The difference between muscle cells of the frog and of the dog is much smaller than the differences between the respective organisms, and this difference becomes even smaller if we look at mitochondria, at myofibrils, or at metabolic pathways and their enzymes. These constitutional constraints are primarily of a qualitative nature but they also comprise quantitative features with regard to the proportions between the parts of the organism. The condition of the fixity of the milieu intérieur, developed by Claude Bernard, is further testimony to the conservative nature of the internal constitution of the body, and at the same time a powerful constraint.

I feel that much of the controversy between physiologists and evolutionary biologists can be resolved if it is recognized that the organism is subjected to these two conflicting constraints between which it must seek a reasonable compromise. Indeed, the organism is at the intersection between the inner and the outer world and, in natural selection, "what determines the success of an individual is the ability of the internal machinery of the organism's body . . . to cope with the challenges of the environment" (Mayr 1991, p. 87). The internal machinery must follow its own rules when the different parts are required to work in concert. And the important point may be that these rules are entirely under the control of the organism as long as the fixity of the milieu intérieur is maintained, and that the adjustments needed are mostly of a quantitative rather than qualitative nature. But problems may arise where the "machinery" forms the actual interface to the outer world with its unpredictabilities and variations that are *not* under the control of the organism. Thus organs of interface, such as the gastrointestinal system, may need to undergo qualitative adaptive changes to cope, for example, with the necessity to feed on indigestible plant fibers. Such adjustments occur between the inner and the outer world; they must be effected with what the organism has available, and it is therefore well conceivable that the solutions found are not necessarily optimal. It is remarkable, however, that such qualitative changes in the organs of interface guarantee considerable independence of the interior conditions of the body: even though animals feed on very different food sources the metabolic processes of the cells all function with the same substrates. Much of modern biology would, by the way, not be possible if the constitution of cells were not so conservative.

When now discussing some of the results of this book I should mention that, in terms of evolution, the range of species considered was rather narrow in that most case studies relate to vertebrates with only a few exceptions. This may be a shortcoming of the debate as it means that we are not discussing major qualitative changes in the body plan – important considerations in evolutionary biology – but rather quantitative differences or shifts in the proportions of the body parts. We are also not looking at an evolutionary time-scale but much of the discussion is based on the comparison of current species. For a large part the papers deal therefore with adjustments within the internal constitution of the body where constitutional constraints may prevail. This limitation must be borne in mind.

## Optimization and symmorphosis revisited

A rather important question for this type of debate is to define what we mean by "optimization" and what criteria are set to test an optimization hypothesis. Some authors in this book argue that optimization is unlikely because a "perfect optimum" cannot be achieved; any deviation from the "line of symmorphosis," it is said, means that the hypothesis must be rejected. I do not believe this is so for several reasons, but most importantly because I regard optimization as a process of becoming, rather than as a state of perfected optimality. This is also what Darwin meant with adaptation. The test for optimization as a process is therefore not whether a particular measurement lies precisely on the line of theoretical optimality, but whether a large number of measurements, obtained under a broad range of variations due to allometry and different adaptations, show a consistent trend towards crowding near the hypothetical optimality line.

Symmorphosis has not been conceived as simply another word for optimal design; in fact, one must wonder whether it crucially needs the stringent definition of optimality; that is, that the design of a given part is precisely made to satisfy its specific function. The primary issue is economic design, enough but not too much, and this particularly with respect to complex functional systems. Here symmorphosis postulates coadjustment of the different links in the chain, or of the different segments in a network, to the needs of serving an overall function. Whether the individual elements are intrinsically "optimally designed" is, in fact, not crucial to this issue. What counts is their contribution to the overall function, and this may, for example, also require safety factors of different magnitude to be built in.

A valid critique was addressed to the proposal that symmorphosis is a hypothesis that can be tested, mainly the suggestion that it is a "null hypothesis" for which one must be able to conclude that the prediction is false. The concept of symmorphosis is indeed too broad to be amenable to a direct test. It should rather be considered a *principle* or theory of coadjusted economic design on the basis of which specific hypotheses can be formulated for the design of individual elements which can then be rigorously tested, and accepted or refuted. Also it appears that the concept can be most productive when applied to complex systems – such as pathways in the form of networks where shared steps and alternative paths are combined – where optimization of all parts is not possible. Here tests based on predictions from symmorphosis may allow the identification of those functions which are dominant in certain parts. The heuristic value of this general hypothesis becomes evident in such instances.

One of the important issues is how to test for symmorphosis. The basic approach is that of determining whether the functional process is limited and whether this is related to design properties (Sih and Gleeson 1995). The first part of the test must be a quantitative physiological study where functional performance is pushed up until a maximum is reached. This may not always be easy. In fact, a true limit can only be measured if the function served by the pathway has alternative means of support: we can measure maximal oxygen consumption only because the muscles can provide the energy to fuel their work by changing to anaerobic glycolysis when oxygen supply is inadequate. The second part of the test is a morphometric study of design parameters which are brought into relation with the limiting functional parameters through a suitable mathematical model.

The test of optimal design or symmorphosis can follow two approaches. First, interventional experiments can be designed by which the balance in a system is disturbed and the response of the system is observed. Some model cases of this are presented in this book: the increase in load on the nutrient supply system by increasing the number of pups of lactating mice (Hammond), or the ablation of half of the lung and its effect on exercise capacity (Hsia). Such experiments test primarily for the malleability of the structures but they can also yield insights into the mechanisms of adjustment. It is evident that the modern tools of molecular genetics offer new possibilities for such interventional experiments which have hitherto not yet been exploited systematically, as pointed out by Sweeney and Feder.

The second approach is to use the methods of comparative physiology in studying adaptations as they occur in experiments of nature, for example the adjustments of structure and function to the differences in body size (cf. Schmidt-Nielsen 1984) or to differences in lifestyle and habitat (cf. Feder *et al.* 1987). These adaptations are to a large part genetically imprinted as a result of natural selection. Their study cannot usually reveal the mechanisms by which adjustment is achieved, or why it is not achieved, but they can form a link to evolutionary biology.

## The evidence scrutinized

For symmorphosis to be acceptable as a basic design principle for optimized coadjustment of structure there must be sufficient evidence that it is upheld in a significant number of cases. This volume is indeed full of case studies where the principle is supported and where one can conclude that design tends to be adjusted to functional needs. If we scrutinize these cases we will find that they are mostly "internal" characteristics of the body which are optimized in relation to conditions which are under the control of the body itself, such as the adjustment of capillary erythrocytes to the mitochondrial oxygen needs, or of the gut epithelium to the nutrient needs of the mammary gland, whereas organs at the interface to the environment appeared as problem cases. This may be a significant differentiation. From the point of view of evolution, Gordon doubts, for example, that the organism can adapt to "resistances to variables of the environment," and this is indeed what we find. So perhaps a distinction must be made between internal elements of functional systems which are coadjusted to requirements entirely under the control of the organism, and elements at the interface to the environment which must adapt both to the needs and constitutional constraints of the body *and* to the stresses or "resistances" of the environment. The adaptation to stresses of the environment is the primary result of natural selection, and it may not always be optimal. If this distinction is made some of the controversies between evolutionists and physiologists may be resolved, as discussed above. But this may also be very significant in a general sense in that the differentiation of organs of interface to meet the challenges of the environment together with the maintenance of an invariant and well-balanced internal milieu affords animals great independence, a freedom to make use of the varied environmental niches for subsistence and existence.

As satisfying as it may be to have found numerous examples where design appears matched to function when variations between species are studied it is perhaps even more fascinating to find cases where the test of structure–function match fails, and to work out the reasons for such maladaptations. We cannot begin to seek this out if we do not pose the question of adaptation or optimization for which we can make precise predictions which can be tested experimentally.

Perhaps one of the most powerful adaptations which natural selection must favor is phenotypic plasticity, the power to adjust or to regulate the quantitative design properties to changing needs of the organism. Phenotypic plasticity results in quantitative malleability of structural design to imposed external or internal stresses; this occurs primarily during ontogenesis and postnatal growth by adjusting the relative growth rates of certain parts, but also in the adult by regulating the morphogenetic events that are associated with the turnover of cells and tissues during maintenance of structure. A few examples should illustrate this. The quantitative distribution of muscle mass is genetically determined by the basic body plan, but it can, secondarily, be adjusted to actual needs. The results of exercise training and of body building, and the converse – the wasting of muscle as a limb is immobilized – are perhaps the most striking examples. The remodeling of bone on altered use and the increase of enzymes for nutrient supply in the intestine when more is needed during lactation and cold exposure, as shown by Hammond, are other examples. The slogan "use it or lose it" is related to this essential property of living things to adjust their make-up to needs or rather to maintain a functional capacity only as long as it is needed. Their overall design is not so much a fixed property as it is the result of a dynamic equilibrium. However, not all organs have this high level of adaptability. Beyond early developmental stages the brain's capacity for structural adaptation is limited, as is well known. As another example, the lung appears to be very limited in its capacity to respond to altered needs in the adult, as shown by the study of Hsia; this may be one of the reasons why the lung needs larger safety factors in the design of its gas exchanger, another reason perhaps being the fact that the lung is an interface organ to the environment where the prevailing conditions for gas exchange are to some extent unpredictable.

Let me, at this point, come back to the conflict between evolutionary biology and physiology. One of the fundamental critiques of symmorphosis, voiced for example by Gordon in his overview, is that it may be more an accident than a principle of adaptation. Accidents are, by definition, rare and haphazard events. The fact that a respectable number of cases have been identified where symmorphosis holds true suggests that this critique may not necessarily stand the test. The chief problem I find with this type of critique is that it almost exclusively emphasizes ultimate causations related to the genetic heritage whereas, in the discussion of quantitative aspects of design, considerations of proximate causations linked partly to phenotypic plasticity may be at least as important.

Finally, one of the provocative approaches to the study of symmorphosis is the study of the design features that characterize species that have reached extremely high-performance levels. Examples are the high-performance athletes such as the Pronghorn antelope, the hummingbirds, and other long-distance migrators, or the smallest mammals, the Etruscan shrew or the Thailand bat. When function is pushed to extremes we would expect design to be as good as it can be – and there is strong evidence that it is.

## A program to be developed

The main purpose of this book was to explore to what extent optimization is a trend in biological design. This cannot have been a definitive account from which to arrive at undisputed conclusions. We must therefore ask where research must go to further approach and eventually solve the riddle of whether symmorphosis is a valid or even important principle governing biological design.

One of the open questions urgently needing further elaboration is the proposition that the organism is bound to find a good – perhaps even optimal – compromise in adjusting to the possibly conflicting external/environmental and internal/constitutional constraints. This will require close and open-minded collaboration between evolutionary biologists, developmental biologists, and physiologists, which can indeed result in a research program of great importance.

With the same goal in mind, we will have to refine and partly redefine the theories of optimization and of symmorphosis and set clear criteria for accepting or rejecting hypotheses derived from these theories. For example, is it indeed necessary that perfect adaptation, optimal design, is achieved, or is it biologically just as important to conclude that there is a strong trend towards optimization under different modes of variation with a result that is perhaps "optimad" rather than perfectly "optimal"? The reason why this differentiation is significant is that biologically it may reveal more about the mechanisms at play in maintaining a well-functioning organism under varying conditions than to refute a theory because perfection is not achieved or not achievable.

In that sense it is also important to explore, in a very intense and concerted effort, the potentials for phenotypic plasticity as a basic mechanism for the adaptation of functional capacity and to attempt to identify the control mechanisms for morphogenesis during growth and maintenance. This can be brought to fruition if the modern means of manipulating organisms by genetic engineering are brought into play in a major effort to establish a well-planned collaboration between physiologists and molecular geneticists.

Such efforts could, indeed, begin to build powerful bridges between evolutionary biology and physiology, and they could eventually resolve the controversies that were at the origin of the debates carried on in this volume.

### Further reading

Boyd, C. A. R. and Noble, D. (1993) *The Logic of Life: The Challenge of Integrative Physiology*. Oxford University Press, Oxford, UK.

Feder, M. E., Bennett, A. F., Burggren, W. W. and Huey, R. B. (1987) *New Directions in Ecological Physiology*. Cambridge University Press, New York, NY.

Gould, S. J. and Lewontin, R. C. (1979) The spandrels of San Marco and the Panglossian paradigm: a critique of the adaptationist programme. *Proc. Royal Society London B*, **205**, 581–98.

Mayr, E. (1991) *One Long Argument: Charles Darwin and the Genesis of Modern Evolutionary Thought*. Harvard University Press, Cambridge, MA.

Schmidt-Nielsen, K. (1984) *Scaling: Why is Animal Size so Important?* Cambridge University Press, New York, NY.

Sih, A. and Gleeson, S. K. (1995) A limits-oriented approach to evolutionary ecology. *TREE*, **10**, 378–82.

Weibel, E. R. (1998) *On Form and Function*. Harvard University Press, Cambridge, MA, in press.

# Index